HeadStart ✓
Primary

THE ALGEBRA BOOK

Written by Laura Sumner

HeadStart ✓
Primary

Acknowledgements:

Author: Laura Sumner
Cover and Page Design: Kathryn Webster

The right of Laura Sumner to be identified as the author of this publication has been asserted by her in accordance with the Copyright, Designs and Patents Act 1998.

HeadStart Primary Ltd
Elker Lane
Clitheroe
BB7 9HZ

T. 01200 423405
E. info@headstartprimary.com
www.headstartprimary.com

Published by HeadStart Primary Ltd 2016 © **HeadStart Primary Ltd 2016**

A record for this book is available from the British Library -
ISBN: 978-1-908767-38-7

HeadStart Primary: THE ALGEBRA BOOK
Introduction

Rationale and book organisation

These questions have been written in line with the objectives from the Mathematics Curriculum.

Algebra is not included as a separate content domain until the Year 6 Programme of Study, when children are expected to use letters to represent unknown numbers. However, the concepts which children need to grasp in order to gain a full understanding of algebraic principles are developed within the mathematics curriculum from Year 1.

Part 1 (page 1 - 36) of this book is based on pre-Year 6 expectations and provides teachers with the opportunity to revise those concepts at an appropriate level for Year 6 pupils. Therefore, it is expected that teachers will use the first part of the book, as necessary, for the whole class or groups of children. It may be that teachers decide to give the earlier pages as homework, prior to beginning the Algebra topic. Alternatively, Part 1 can be used in earlier year groups in preparation for Year 6, as appropriate.

Part 2 is based on the Year 6 expectations and includes questions to reflect all the Year 6 objectives from the Mathematics Curriculum Algebra content domain. In addition, 'using the correct order of operations' (taken from the Year 6 Number – addition, subtraction, multiplication and division domain) and 'solving equations with an unknown number on both sides' (challenge beyond that expected within the Year 6 curriculum) is applied to algebra.

Approaches to teaching algebra

In order to support the teaching of algebraic concepts, reminders of appropriate strategies, which model important principles in algebra, are identified on some pages. It is understood that schools may vary their teaching approach from the model shown. In such cases, teachers should substitute the appropriate school model.

When multiplying in algebra, common practice is to omit the multiplication symbol. However, so that children can focus on the concept of letters representing missing numbers, the early pages within the Year 6 Expectations do include the symbol. Later pages introduce the practice of omitting the multiplication symbol.

Differentiation

In general, the questions are arranged so that they become progressively more difficult on each page. Additionally, for each objective, there is a page or pages which give the opportunity to practise the objective at a standard level (relatively simple calculations e.g. smaller numbers, limited crossing of the tens boundary etc.), as well as a parallel page or pages at a more challenging level (more complex calculations, larger numbers etc.). This is indicated by (ch) at the bottom right hand corner of the page. Consequently, to aid differentiation, those children who need to focus purely on the algebraic concepts are able to work on relatively simple calculation aspects, whilst those children who are capable of understanding algebra with more challenging calculations are also provided for. It may be appropriate for some children to attempt the simpler pages before moving on to the challenging pages. The last objective 'Solve equations with an unknown number on both sides', only contains challenging pages as the concept covers challenge beyond that expected within the Year 6 curriculum.

Using the worksheets

The book is designed so that children are able to write answers on the photocopied sheets, which may be particularly useful if given as homework. However, should it be more appropriate, pupils can easily transcribe the work into exercise books. Where substantial 'working out' needs to be completed, this may have to be carried out in exercise books or on separate paper.

CONTENTS

Part 1: Pre-Year 6 Expectations

CONTENTS

Part 2: Year 6 (and beyond) Expectations

CONTENTS

Year 6 Expectations

Part 1

Pre-Year 6 Expectations

This section will help you prepare to study algebra!

Name .... W37 .... Class ... 26-5-24 .... Date ....

Use the relationship between addition and subtraction to solve missing number problems.

1 Complete the following.

a [5] + 4 = 9

b 6 + [7] = 13

c 26 + [61] = 87

d [43] + 33 = 76

e 146 + [113] = 259

f [112] + 222 = 334

---

2 Now try these.

a 15 – [4] = 11

b [19] – 3 = 16

c 32 – [8] = 24

d [99] – 42 = 57

e 174 – [112] = 62

f [195] – 85 = 110

Checked

3 Have a go at these.

a 14 + [62] = 76

b [88] – 37 = 51

c 23 + [61] = 84

d [178] – 121 = 57

e 232 + [264] = 496

f [387] – 236 = 151

Use the relationship between addition and subtraction to solve missing number problems.

1 Complete the following.

a) $\boxed{54}$ + 42 = 96

b) 123 + $\boxed{264}$ = 387

c) $\boxed{52}$ + 375 = 427

d) 124 + $\boxed{79}$ = 203

e) $\boxed{182}$ + 752 = 934

f) 384 + $\boxed{1253}$ = 1637

2 Now try these.

a) $\boxed{79}$ − 53 = 26

b) 168 − $\boxed{64}$ = 104

c) $\boxed{203}$ − 69 = 134

d) $\boxed{136}$ − 89 = 47

e) 7084 − $\boxed{6841}$ = 243

f) $\boxed{2374}$ − 2085 = 289

3 Have a go at these.

a) 156 + $\boxed{222}$ = 378

b) $\boxed{280}$ − 157 = 123

c) $\boxed{251}$ + 285 = 536

d) $\boxed{424}$ − 246 = 178

e) 637 + $\boxed{1082}$ = 1719

f) $\boxed{4726}$ − 1187 = 3539

(ch) ALGEBRA

Name .... W 37 .... Class .... Date 26·5·24

Use the relationship between addition and subtraction to solve missing number problems.

1 Complete the following by filling in all the boxes with the numbers from the first equation. An example is shown.

checked

a

$$13 \quad + \quad 54 \quad = \quad 67$$

so $\boxed{54}$ + $\boxed{13}$ = 67

and 67 − $\boxed{54}$ = $\boxed{13}$

and $\boxed{67}$ − $\boxed{13}$ = $\boxed{54}$

b

$$85 \quad - \quad 33 \quad = \quad 52$$

so 85 − $\boxed{52}$ = $\boxed{33}$

and $\boxed{33}$ + $\boxed{52}$ = 85

and $\boxed{52}$ + $\boxed{33}$ = $\boxed{85}$

c

$$144 \quad + \quad 24 \quad = \quad 168$$

so 168 − $\boxed{144}$ = $\boxed{24}$

and $\boxed{24}$ + $\boxed{144}$ = $\boxed{168}$

and $\boxed{168}$ − $\boxed{24}$ = $\boxed{144}$

d

$$127 \quad - \quad 83 \quad = \quad 44$$

so $\boxed{83}$ + $\boxed{44}$ = 127

and $\boxed{44}$ + $\boxed{83}$ = $\boxed{127}$

and $\boxed{127}$ − $\boxed{44}$ = $\boxed{83}$

3

Continued overleaf **ALGEBRA**

checked

e   423   +   68   =   491

so [68] + [423] = [491]

and [491] − [68] = [423]

and [491] − [423] = [68]

f   423   −   141   =   282

so [141] + [282] = [423]

and [423] − [282] = [141]

and [282] + [141] = [423]

g   306   +   124   =   430

so [430] − [124] = [306]

and [430] − [306] = [124]

and [124] + [306] = [430]

h   [588]   −   431   =   157

so [431] + [157] = [588]

and [588] − [157] = [431]

and [157] + [431] = [588]

You're doing really well!

4 ✓

ALGEBRA

Name... Class........................ Date 2-6-24

Use the relationship between addition and subtraction to solve missing number problems.

1 Complete the following by filling in all the boxes with the numbers from the first equation. An example is shown.

a *Checked*

	146	+	58	=	204
so	58	+	146	=	204
and	204	–	146	=	58
and	204	–	58	=	146

b

	165	–	63	=	102
so	165	–	102	=	63
and	63	+	102	=	165
and	102	+	63	=	165

c

	432	+	51	=	483
so	483	–	432	=	51
and	51	+	432	=	483
and	483	–	51	=	432

d

	364	–	153	=	211
so	153	+	211	=	364
and	211	+	153	=	364
and	364	–	211	=	153

e

356 + 506 = 862

so 506 + 356 = 862

and 862 − 506 = 356

and 862 − 356 = 506

f

722 − 336 = 386

so 336 + 386 = 722

and 722 − 386 = 336

and 386 + 336 = 722

g

4723 + 276 = 4999

so 4999 − 276 = 4723

and 4999 − 4723 = 276

and 276 + 4723 = 4999

h

6032 − 1382 = 4650

so 1382 + 4650 = 6032

and 4650 + 1382 = 6032

and 6032 − 4650 = 1382

Checked

Well done!

ch ALGEBRA

Name .... Class .... Date ....
2.6-24

Use the relationship between addition and subtraction to solve missing number word problems.

---

1 Read each problem below. **Put a circle** around the equation you would use to solve the problem.

a There are 56 blue and red cars in the garage. 23 are blue. How many red cars are there?

Checked

$$23 \ - \ 56 \ = \ \boxed{?}$$

$$56 \ - \ 23 \ = \ \boxed{?}$$

$$\boxed{?} \ - \ 23 \ = \ 56$$

$$56 \ + \ 23 \ = \ \boxed{?}$$

---

b Mrs Shah baked 86 cakes for her daughter's birthday party. After the party, 25 were left. How many cakes were eaten at the party?

$$86 \ + \ 25 \ = \ \boxed{?}$$

$$25 \ - \ 86 \ = \ \boxed{?}$$

$$86 \ - \ 25 \ = \ \boxed{?}$$

$$\boxed{?} \ - \ 25 \ = \ 86$$

---

c From Monday to Thursday, Esme collected 43 house points. On Friday she gained another 14. How many house points did Esme collect altogether?

$$43 \ - \ 14 \ = \ \boxed{?}$$

$$43 \ + \ 14 \ = \ \boxed{?}$$

$$43 \ - \ \boxed{?} \ = \ 14$$

$$14 \ + \ \boxed{?} \ = \ 43$$

Continued overleaf ALGEBRA

Name.. Class........................ Date......................

9-6-24

Checked

2 Now try this

a Before lunch, Ruthie drove 56 miles. After lunch she drove 27 miles. How many miles had she driven altogether?

56 + ? = 27   56 + 27 = ?

27 − 56 = ?   56 − 27 = ?

---

b Use the equation you have chosen to find the answer to the problem.

83

---

3 Have a go at this one.

a Tom got his spending money on Saturday. He spent £3.50 at the shop and had £5.00 left. How much spending money had Tom received?

£3.50 + £5.00 = ?   £5.00 + ? = £3.50

£3.50 − £5.00 = ?   £5.00 − £3.50 = ?

---

b Use the equation you have chosen to find the answer to the problem.

£8.50

---

4 For these problems, write down an equation you could use to solve the problem. Then use your equation to find the answer.

a 65 people were on the bus. Some more people got on at Greentree, but no one got off. Then there were 82 people on the bus. How many people got on at Greentree.

   =

You can use this space for your working.

b A shop had 79 iPads. On Monday it sold 23 iPads. How many iPads did the shop have left?

   =

You can use this space for your working.

c The classroom library had 126 books. 42 were out on loan. How many books were on the library book shelves?

   =

You can use this space for your working.

ALGEBRA

Name .... _w 39_ .... Class .... _checked_ .... Date ....

Use the relationship between addition or subtraction to solve missing number word problems.

---

1 Read each problem below. **Put a circle around** the equation you would use to solve the problem.

a Afzal had 182 football cards altogether. 73 had a picture of a premiership player. How many cards did not have a picture of a permiership player?

$$73 + 182 = \boxed{?}$$

$$182 - 73 = \boxed{?}$$

$$73 - 182 = \boxed{?}$$

$$182 + 73 = \boxed{?}$$

---

b Mr Bread, the baker, sold 473 loaves, on Monday. On Tuesday he sold 189 loaves. How many loaves did Mr Bread sell altogether?

$$473 - 189 = \boxed{?}$$

$$189 + \boxed{?} = 473$$

$$473 + 189 = \boxed{?}$$

$$189 - 473 = \boxed{?}$$

---

c Jan's netball team scored 256 goals during the season. They had 79 goals scored against them. What was the difference between the goals for and against?

$$79 - 256 = \boxed{?}$$

$$256 + 79 = \boxed{?}$$

$$79 + 256 = \boxed{?}$$

$$256 - 79 = \boxed{?}$$

Name........W 39 Checked........... Class..................... Date.........................

2 Now try this

a Janey's family had to drive 1652 kilometres to their holiday destination. On the first day, they drove 967 kilometres. How much further did they have to drive?

1652 + 967 = **?**   1652 – 967 = **?**

1652 + **?** = 967   967 – 1652 = **?**

- -

b Use the equation you have chosen to find the answer to the problem.

$$\begin{array}{r} 1652 \\ -\ 967 \\ \hline 685 \end{array}$$

685

3 Have a go at this one.

a 1908 pupils attend Sunnyside High School. 1184 are boys. How many are girls?

1184 – 1908 = **?**   1908 + 1184 = **?**

1908 + **?** = 1184   1908 – 1184 = **?**

- -

b Use the equation you have chosen to find the answer to the problem.

$$\begin{array}{r} 1908 \\ -1184 \\ \hline 724 \end{array} \qquad \begin{array}{r} 1980 \\ -1184 \\ \hline 796 \end{array}$$

724

Continued overleaf **ch** **ALGEBRA**

Name .... W39 .... Class .... Checked .... Date ....

4 For these problems, write down an equation you could use to solve the problem. Then use your equation to find the answer.

a Anya was saving up for a laptop. She saved £375.50 from her earnings and £50.78 from her birthday money. How much did Anya save altogether?

£ 375.50   [+]   £ 50.78   =   £ 426.28

You can use this space for your working.

```
  5078
37550
42628
```

b The perimeter of a garden measured 156.35m. There was a fence aound 78.5 metres of the perimeter. How much of the garden's perimeter did not have a fence?

156.35   [−]   78.5   =   77.85

You can use this space for your working.

```
156.35
 78.50
 77.85
```

c To arrive at their holiday destination, Julian's family drove 87km before lunch and 87km after lunch. How far did they drive altogether?

87   [X]   2   =   174

You can use this space for your working.

```
87
 2
174
```

ch ALGEBRA

Name .... W40   Class .... 15-6-24 .... Date ....

Checked

Use the relationship between multiplication and division to solve missing number problems.

1 Complete the following.

a [6] x 4 = 24   d [6] x 7 = 42

b 6 x [5] = 30   e 10 x [15] = 150

c 5 x [12] = 60   f 8 x [12] = 96

---

2 Now try these.

a [27] ÷ 3 = 9   d [35] ÷ 7 = 5

b [24] ÷ 8 = 3   e 84 ÷ [12] = 7

c 48 ÷ [8] = 6   f [72] ÷ 9 = 8

---

3 Have a go at these.

a [8] x 5 = 40   d [260] ÷ 10 = 26

b [42] ÷ 6 = 7   e 9 x [12] = 108

c 6 x [9] = 54   f [98] ÷ 7 = 14

ALGEBRA

Name.. Class....................... Date........................

W40   15-6-24

checked

Use the relationship between multiplication and division to solve missing number problems.

1 Complete the following.

a [7] x 12 = 84

d 25 x [18] = 450

b 11 x [13] = 143

e [52] x 14 = 728

c [12] x 15 = 180

f 23 x [30] = 690

050
230690
62
x

2 Now try these.

a [96] ÷ 12 = 8

d 180 ÷ [4] = 45

b 165 ÷ [11] = 15

e [1008] ÷ 36 = 28

c [2280] ÷ 30 = 76

3
228

f [4823] ÷ 53 = 91

3 Have a go at these.

1056

a [31] x 30 = 930

d [105x] ÷ 22 = 48

b [966] ÷ 42 = 23

e 38 x [16] = 608

42
23
c 35 x [12] = 420   126
840
966

012
35/120
35
70

f [1512] ÷ 72 = 21

72
21
72
1440
1512

ch ALGEBRA

Name... Class....................... Date........................

Use the relationship between multiplication and division to solve missing number problems.

1 Complete the following by filling in all the boxes with the numbers from the first equation. An example is shown.

a

| | | 8 | X | 7 | = | 56 |

so [7] X [8] = 56

and 56 ÷ [7] = [8]

and [56] ÷ [8] = [7]

b

54 ÷ 9 = 6

so 54 ÷ [6] = 9

and [6] X [9] = 54

and [9] X [6] = [54]

c

12 X 8 = 96

so 96 ÷ [8] = [12]

and [8] X [12] = [96]

and [96] ÷ [12] = [8]

d

78 ÷ 6 = 13

so [13] X [6] = 76

and [6] X [13] = 76

and [78] ÷ [13] = [6]

15

Continued overleaf **ALGEBRA**

e   13   x   15   =   195

so | 15 | x | 13 | = | 195 |

and | 195 | ÷ | 15 | = | 13 |

and | 195 | ÷ | 13 | = | 15 |

f   153   ÷   9   =   17

so | 9 | x | 17 | = | 153 |

and | 153 | ÷ | 17 | = | 9 |

and | 17 | x | 9 | = | 153 |

g   23   x   12   =   276

Checking

so | 276 | ÷ | 23 | = | 12 |

and | 276 | ÷ | 12 | = | 23 |

and | 12 | x | 23 | = | 276 |

h | 72 | ÷   8   =   9

so | 9 | x | 8 | = | 72 |

and | 72 | ÷ | 9 | = | 8 |

and | 8 | x | 9 | = | 72 |

You're doing a great job!

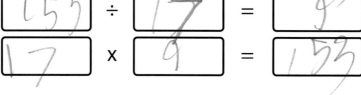

Name.. Class....................... Date.........................

Use the relationship between multiplication and division to solve missing number problems.

1 Complete the following by filling in all the boxes with the numbers from the first equation. An example is shown.

a

13　　X　　21　　=　　273

so | 21 | X | 13 | = | 273

and | 273 | ÷ | 13 | = | 21

and | 273 | ÷ | 21 | = | 13

b

322　　÷　　14　　=　　23

so | 322 | ÷ | 23 | = | 14

and | 14 | X | 23 | = | 322

and | 23 | X | 14 | = | 322

c

14　　X　　16　　=　　224

so | 224 | ÷ | 16 | = | 14

and | 16 | X | 14 | = | 224

and | 224 | ÷ | 14 | = | 16

d

345　　÷　　23　　=　　15

so | 23 | X | 15 | = | 345

and | 15 | X | 23 | = | 345

and | 345 | ÷ | 15 | = | 23

17 *Continued overleaf* **ch** **ALGEBRA**

Checked 23-6-20

e

17   x   [26]   =   442

so   [26]   x   [17]   =   [442]

and   [442]   ÷   [26]   =   [17]

and   [442]   ÷   [17]   =   [26]

f

[1352]   ÷   26   =   52

so   [52]   x   [26]   =   [1352]

and   [1345]   ÷   [52]   =   [26]

and   [26]   x   [52]   =   [1352]

g

[25]   x   32   =   800

so   [800]   ÷   [32]   =   [25]

and   [32]   x   [25]   =   [800]

and   [800]   ÷   [25]   =   [32]

52
36
26

312
1040

17
12
34
170
204

h

[1176]   ÷   56   =   21

so   [56]   x   [21]   =   [1176]

and   [56]   x   [21]   =   [1176]

and   [1176]   ÷   [21]   =   [56]

1352
56
21

56
1120
1176

17
6
102

32
15
160
320
480

Keep up the good work!

ch **ALGEBRA**

Name.. Class........................ Date.......................

Use multiplication or division to solve missing number word problems.

1 Read each problem below. **Put a circle** around the equation you would use to solve the problem.

a Baz, Gracie and Shazia share £27 evenly between themselves. How much does Gracie get?

£23 x 3 = **?**   3 x £27 = **?**

£27 ÷ 3 = **?**   3 ÷ £27 = **?**

Checked

---

b In the reception class, there are ten books in each reading box. There are eight reading boxes. How many books are there altogether?

10 ÷ 8 = **?**   10 x 8 = **?**

10 x **?** = 8   8 ÷ 10 = **?**

---

c Six friends had £2.50 each to go to the fair. How much had they altogether?

6 x **?** = £2.50   £2.50 ÷ 6 = **?**

£2.50 x 6 = **?**   6 ÷ £2.50 = **?**

2 Now try this

a Katya thought of a number and divided it by 9. Her answer was 12. What was Katya's number?

12 x **?** = 9   9 x 12 = **?**

12 ÷ 9 = **?**   9 x **?** = 12

- -

b Use the equation you have chosen to find the answer to the problem.

108

- -

3 Have a go at this one.

a Amir had the same number of marbles in each of 8 tubs. He had 56 marbles altogether. How many marbles were in one tub?

56 x 8 = **?**   56 x **?** = 8

56 ÷ 8 = **?**   8 ÷ **?** = 56

- -

b Use the equation you have chosen to find the answer to the problem.

7

Name .... WU2 2-7-24 Class .... Date ....

4 For these problems, write down an equation you could use to solve the problem. Then use your equation to find the answer.

a Six friends each gave the same amount to the school charity fund. They gave £12 altogether. How much did they each give?

| 12 | ÷ | 6 | = | 2 |

checked

You can use this space for your working.

---

b Laura, Riz and Chloe each swam 23 lengths of the pool. How many lengths did they swim altogether?

| 23 | × | 3 | = | 69 |

You can use this space for your working.

---

c Brenda's Bakery sold 27 loaves of bread every day for 6 days. How many loaves did they sell altogether?

| 27 | × | 6 | = | 162 |

You can use this space for your working.

$$\begin{array}{r} \overset{4}{2}7 \\ 6 \\ \hline 162 \end{array}$$

ALGEBRA

Name.................... WUB a-7-24 Class.................... Date....................

Use multiplication or division to solve missing number word problems.

1 Read each problem below. **Put a circle** around the equation you would use to solve the problem.

a Seven people bought a ticket for a fairground ride. They paid £21.70 altogether. How much did each ticket cost?

£21.70 x 7 = ?   7 ÷ £21.70 = ?

£21.70 ÷ 7 = ?   7 x £21.70 = ?

---

b Archie saves £1.50 per week for 56 weeks. How much did he save altogether?

? x £1.50 = 56   £1.50 ÷ 56 = ?

56 ÷ £1.50 = ?   £1.50 x 56 = ?

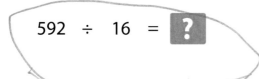

---

c A shop sells 592 bars of chocolate over 16 days. What is the average number of bars sold per day?

592 ÷ 16 = ?   592 x ? = 16

592 x 16 = ?   16 x ? = 592

Name .... Class .... Date ....

W43
9-7-24

2 Now try this

a Asmat thought of a number and divided it by 18. His answer was 21. What was Asmat's number?

21 x 18 = ?   21 ÷ 18 = ?

18 x ? = 21   21 ÷ ? = 18

- -

b Use the equation you have chosen to find the answer to the problem.

Checked
18
21
18
260
378

378

3 Have a go at this one.

a The pasta factory divided pasta between 56 bags with 55 grams in each bag. How much pasta was there to start with?

55g x ? = 56   56 ÷ 55g = ?

55g ÷ 56 = ?   56 x 55g = ?

- -

b Use the equation you have chosen to find the answer to the problem.

56
55
280
2800
3080

3080 g

Continued overleaf (ch) ALGEBRA

Name .... *W43* 9-7-24 .... Class .... Date ....

4 For these problems, write down an equation you could use to solve the problem. Then use your equation to find the answer.

a A box can hold 12 biscuits. How many boxes are needed to hold 768 biscuits?

| 768 | ÷ | 12 | = | 64 |

- -

You can use this space for your working.

064
12)768

b Mo's teacher asked, "If a number divided by 32 is 98, what is the number?". What should Mo's answer have been?

| 32 | ✗ | 98 | = | 3136 |

- -

You can use this space for your working.

98
32

196
2940

c An average of 872 people attended the local football team's matches over 22 games, during a season. How many people attended altogether?

| 872 | ✗ | 22 | = | 19184 |

- -

You can use this space for your working.

checked

872
22
1744
17440
19184

© Copyright HeadStart Primary Ltd 2016   24   (ch) ALGEBRA

Understand that any symbol can be used to represent numbers.

KEY

 = 6   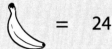 = 48   🍌 = 24

🍎 = 15   🍐 = 5   🐸 = 54

Checked

1 Use the key above to solve the following.

a   + = $\boxed{21}$ c   + = $\boxed{69}$

b 🍌 − 🍐 = $\boxed{19}$ d 🐸 − 🍌 = $\boxed{30}$

---

2 Now try these.

a   x = $\boxed{30}$ c   x = $\boxed{75}$

b   ÷ = $\boxed{3}$ d   ÷ = $\boxed{4}$

---

3 Have a go at these.

a   + = $\boxed{72}$ e   − = $\boxed{39}$

b   − = $\boxed{33}$ f   x = $\boxed{288}$

c   x 🍌 = $\boxed{144}$ g   + 🐸 = $\boxed{102}$

d 🐸 ÷  = $\boxed{4}$ h   ÷ = $\boxed{8}$

2
24
6
144
4
48
6
288

Name....................W44..............5-7-24... Class........................ Date........................

Understand that any symbol can be used to represent numbers.

KEY

 = 9   = 105   = 135

= 72   = 15   = 270

1 Use the key above to solve the following.

a + = 24   c + = 87

b −  = 63   d − = 163

2 Now try these.

a X = 135 ~~228~~   c X = 648

b ÷ = 8   d  ÷ = 7

3 Have a go at these.

a +  = 207   e 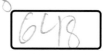 − = 198

b − = 90   f X  = 9720

c X = 2430   g + = 342

d ÷ = 9   h ÷ = 30

ch ALGEBRA

Name .... Class .... Date ....

Understand that any symbol can be used to represent numbers. Use the relationships between operations.

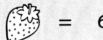 = 6   = 48   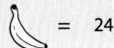 = 24

 = 15   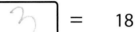 = 5   = 54

1 Use the key above to solve the following.

a + [3] = 18   **c**  + [42] = 96

b [23] − = 18   **d** [88] − = 40

2 Now try these.

a ÷ [8] = 6   **c** x [12] = 72

b [9] x = 45   **d** [90] ÷ = 6

3 Have a go at these.

a [72] + = 96   **e** x [2] = 48

b − [48] = 6   **f** [79] − = 25

c [11] x = 55   **g** + [100] = 148

d 42 ÷ [7] =   **h** [105] ÷ = 7

Name .... W44 ....   Class .... 15-7-24 ....   Date ....

Understand that any symbol can be used to represent numbers. Use the relationships between operations.

Checked

KEY

 = 9

 = 105

= 135

= 72

 = 15

= 270

1 Use the key above to solve the following.

a  + $\boxed{56}$ = 65

c   + $\boxed{55}$ = 127

b $\boxed{87}$ − = 72

d $\boxed{346}$ − = 241

2 Now try these.

a   ÷ $\boxed{3}$ = 45

c   x $\boxed{6}$ = 630

b   x $\boxed{13}$ = 195

d $\boxed{936}$ ÷ = 13

3 Have a go at these.

a $\boxed{69}$ + = 204

e   x $\boxed{463}$ = 6945

b $\boxed{1082}$ − = 812

f $\boxed{1134}$ − 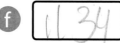 = 864

c   x $\boxed{12}$ = 864

g   + $\boxed{2092}$ = 2164

d $\boxed{6615}$ ÷ 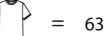 = 63

h $\boxed{8715}$ ÷ 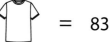 = 83

$\begin{array}{r} 72 \\ 2 \\ \hline 144 \end{array}$ 72

ch ALGEBRA

Make both sides of an equation equal, where one side contains a missing number

1 Complete the following.

a) $4 \quad + \quad \boxed{17} \quad = \quad 9 \quad + \quad 12$

b) $27 \quad - \quad 9 \quad = \quad \boxed{22} \quad - \quad 4$

c) $4 \quad \times \quad \boxed{6} \quad = \quad 2 \quad \times \quad 12$

d) $42 \quad \div \quad 7 \quad = \quad \boxed{12} \quad \div \quad 2$

e) $12 \quad + \quad 23 \quad = \quad \boxed{20} \quad + \quad 15$

f) $69 \quad - \quad \boxed{24} \quad = \quad 57 \quad - \quad 12$

g) $81 \quad \div \quad \boxed{9} \quad = \quad 54 \quad \div \quad 6$

h) $8 \quad \times \quad 13 \quad = \quad \boxed{52} \quad \times \quad 2$

i) $66 \quad \div \quad \boxed{11} \quad = \quad 72 \quad \div \quad 12$

j) $46 \quad + \quad 32 \quad = \quad 24 \quad + \quad \boxed{54}$

k) $14 \quad \times \quad 8 \quad = \quad \boxed{16} \quad \times \quad 7$

l) $\boxed{66} \quad - \quad 28 \quad = \quad 72 \quad - \quad 34$

2 Now try these.

23-7-24

Checked

a 26 + [8] = 44 − 10

b 49 − [7] = 21 + 21

c 3 x 2 = [30] ÷ 5

d 96 ÷ 8 = [3] x 4

e [12] + 15 = 34 − 7

f 29 + 33 ~~32~~ = [77] − 15

g 2 x [4] = 72 ÷ 9

h [392] ÷ 7 = 2 x 28

56

63

4
56
7
___
392

i 82 − [19] = 39 + 24

j 4 x [4] = 160 ÷ 10

k 157 − 42 = [15] + 100

l [120] ÷ 4 = 6 x 5

3 Have a go at these. They may be tricky - look carefully at the operations!

a 8 + [22] = 6 x 5

b 54 ÷ 9 = [29] – 23

c 7 x [4] = 37 – 9

d 7 + 5 = [24] ÷ 2

e [9] x 7 = 49 + 14

f 84 – [67] = 13 + 4

g 64 ÷ [4] = 7 + 9

h 96 ÷ 3 = 17 + [15]

i 84 x 3 = 102 + [150]

j 29 + 17 = 92 ÷ [2]

k 466 – [222] = 4 x 61

l [648] ÷ 9 = 18 x 4

Name..W4b 23-7-24..................................... Class...................... Date.......................

4 Now try these. Think carefully.

a 23 + 49 = 8 x [9] = 104 − 32

72

b [93] − 69 = 96 ÷ 4 = 12 x 2

c 105 ÷ 3 = 7 x 5 = 18 + [17]

d 360 − 296 = [512] ÷ 8 = 25 + 39

64

e 56 + 32 = [88] x 11 = 176 ÷ 2 = [102] − 14

88

f [84] + 79 = 8 x 21 = [168] ÷ 1 = 392 − 224

138

g 6 x 23 = 150 − [12] = 276 ÷ [2] = 60 + 78

h 45 + [105] = 450 ÷ 3 = 6 x [25] = 232 − 82

Making good progress!

32

ALGEBRA

Make both sides of an equation equal, where one side contains a missing number

1 Complete the following by putting the missing numbers in the boxes.

a) $26 + \boxed{29} = 21 + 34$

b) $54 - 27 = \boxed{65} - 38$

c) $12 \times \boxed{13} = 39 \times 4$

d) $198 \div 11 = \boxed{108} \div 6$

e) $106 + 49 = 76 + \boxed{79}$

f) $113 - 37 = \boxed{63} - 87$

g) $364 \div \boxed{14} = 338 \div 13$

h) $34 \times 23 = 17 \times \boxed{46}$

i) $143 + 268 = \boxed{84} + 327$

j) $391 \div \boxed{23} = 578 \div 34$

k) $5.2 \times 35 = 8 \times \boxed{22.75}$

l) $\boxed{529.5} - 246 = 356.5 - 78$

Name... Class............................ Date.......................

2 Complete the following by putting the missing numbers in the boxes.

a) $53 + \boxed{62} = 193 - 78$

b) $3576 - \boxed{2394} = 323 + 859$

c) $26 \times 24 = \boxed{8112} \div 13$

d) $\boxed{16} \times 1.5 = 432 \div 18$

e) $\boxed{33.2} + 15.7 = 63.8 - 14.9$

f) $56.4 + 32.5 = \boxed{124.9} - 36$

g) $\boxed{1.25} \times 3.6 = 81 \div 18$

h) $\boxed{5880} \div 5 = 98 \times 12$

i) $1643 - \boxed{694.2} = 592 + 356.8$

j) $27 \times \boxed{3.5} = 1134 \div 12$

k) $6049 - 32.42 = \boxed{122.58} + 5894$

l) $\boxed{336} \div 3.5 = 3.84 \times 25$

34

Continued overleaf **ch** **ALGEBRA**

3 Have a go at these. They may be tricky - look carefully at the operations.

a 98 + [62] = 64 x 2.5

b 630 ÷ 18 = [199.5] − 164.5

c 4.5 x [62] = 1064 − 785

d 1063 + 2458 = [31689] ÷ 9

e [56] x 2.7 = 45.25 + 105.95

f 742.6 − [90.46] = 36.23 x 18

g 408 ÷ [17] = 6.38 + 17.62

h 2470 ÷ 26 = 72.68 + [822.32]

i 1262 x 18 = 9852 + [12864]

j 8.423 + 39.577 = 1680 ÷ [35]

k 82291 − [5907] = 1364 x 56

l [491.13] ÷ 2.7 = 1006 − 824.1

Continued overleaf **ch** **ALGEBRA**

4 Now try these. Think carefully.

a   156   +   456   =   18   x   [34]   =   1242   −   630

b   [6384]   −   6369   =   36   ÷   2.4   =   12   x   1.25

c   1352   ÷   26   =   16   x   3.25   =   36.23   +   [15.77]

d   857.3   −   789.3   =   [285.6]   ÷   4.2   =   36.248   +   31.752

e   23.29 + 13.71 =   [2]   x 18.5 = 240.5 ÷ 6.5 =   [93.64]   − 56.64

f   [34.36]   + 52.31 = 27 x 3.21 =   [953.37]   ÷ 11 = 130.17 − 43.5

You're doing well!

Part 2

Year 6 (and beyond) Expectations

You are now ready to study algebra and practise your skills!

Name .... Class .... Date ....

Begin to use letters to represent variables.

1 Put the missing numbers in the boxes.

a $\boxed{}$ + 33 = 49

c  x 6 = 48

 = $\boxed{8}$

b 56 − $\boxed{}$ = 4

d 128 − 🐸 = 14

🐸 = $\boxed{}$

---

2 In algebra, instead of symbols to represent numbers, we can use letters.

a x + 26 = 54

x = $\boxed{}$

e 48 + y = 92

y = $\boxed{}$

- - - - - - - - - - - - - - - - - - - -

b 27 − y = 19

y = $\boxed{}$

f x − 24 = 48

x = $\boxed{}$

- - - - - - - - - - - - - - - - - - - -

c 42 + a = 68

a = $\boxed{}$

g b + 47 = 152

b = $\boxed{}$

- - - - - - - - - - - - - - - - - - - -

d b − 27 = 62

b = $\boxed{}$

h 183 − a = 127

a = $\boxed{}$

Continued overleaf **ALGEBRA**

3 Now try these.

a x × 5 = 55

x = ☐

b 36 ÷ y = 4

y = ☐

c 9 × a = 45

a = ☐

d b ÷ 7 = 6

b = ☐

e 12 × y = 72

y = ☐

f x ÷ 11 = 11

x = ☐

g b × 8 = 96

b = ☐

h 108 ÷ a = 12

a = ☐

4 Have a go at these. Take care - the operations are mixed up.

a 68 + x = 94

x = ☐

b y − 52 = 39

y = ☐

c 9 × a = 108

a = ☐

d b ÷ 3 = 32

b = ☐

e y × 7 = 56

y = ☐

f 184 − x = 141

x = ☐

g b + 27 = 94

b = ☐

h 96 ÷ a = 8

a = ☐

ALGEBRA

Name .... Class .... Date ....

Begin to use letters to represent variables.

1 Put the missing numbers in the boxes.

a ☐ + 43 = 126

c 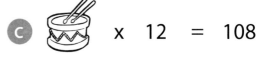 x 12 = 108

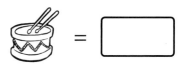 = ☐

b 108 ÷ ☐ = 6

d 374 − 🎎 = 87

🎎 = ☐

---

2 In algebra, instead of symbols to represent numbers, we can use letters.

a x + 159 = 264

x = ☐

e 487 + y = 726

y = ☐

b 232 − y = 87

y = ☐

f x − 324 = 287

x = ☐

c 118 + a = 423

a = ☐

g b + 267 = 1486

b = ☐

d b − 157 = 348

b = ☐

h 1462 − a = 744

a = ☐

Continued overleaf (ch) **ALGEBRA**

Name .... Class .... Date ....

3 Now try these.

a $x \times 12 = 228$

$x = \boxed{}$

b $195 \div y = 13$

$y = \boxed{}$

c $9 \times a = 234$

$a = \boxed{}$

d $b \div 14 = 53$

$b = \boxed{}$

e $17 \times y = 646$

$y = \boxed{}$

f $840 \div x = 21$

$x = \boxed{}$

g $b \times 24 = 1272$

$b = \boxed{}$

h $a \div 36 = 59$

$a = \boxed{}$

4 Have a go at these. Take care - the operations are mixed up.

a $152 + x = 932$

$x = \boxed{}$

b $y - 488 = 229$

$y = \boxed{}$

c $27 \times a = 999$

$a = \boxed{}$

d $b \div 23 = 831$

$b = \boxed{}$

e $y \times 36 = 180$

$y = \boxed{}$

f $123 - x = 84.2$

$x = \boxed{}$

g $b + 463.2 = 541.8$

$b = \boxed{}$

h $2688 \div a = 32$

$a = \boxed{}$

Name .... Class .... Date ....

Solve equations where letters represent missing numbers.

Remember, in algebra we can use letters to represent missing numbers.

1 Find the value of x or y in these equations. Put your answers in the boxes.

a $x + 3 = 14$

$x =$ ⬚

b $7 + y = 18$

$y =$ ⬚

c $8 + y = 22$

$y =$ ⬚

d $x + 7 = 34$

$x =$ ⬚

2 Now find the value of a or b in these.

a $6 - a = 2$

$a =$ ⬚

b $b - 4 = 12$

$b =$ ⬚

c $b - 7 = 9$

$b =$ ⬚

d $17 - a = 8$

$a =$ ⬚

3 Find the value of the letters in these mixed addition and subtraction equations.

a $14 + x = 18$

$x =$ ⬚

b $y - 6 = 37$

$y =$ ⬚

c $38 - b = 24$

$b =$ ⬚

d $y + 64 = 85$

$y =$ ⬚

e $a - 9 = 87$

$a =$ ⬚

f $55 + x = 92$

$x =$ ⬚

4 Now try to find a or b in these equations.

a $\quad a \;\times\; 3 \;=\; 24$

$\qquad a \;=\; \boxed{}$

c $\quad 7 \;\times\; b \;=\; 35$

$\qquad b \;=\; \boxed{}$

b $\quad 4 \;\times\; b \;=\; 32$

$\qquad b \;=\; \boxed{}$

d $\quad a \;\times\; 9 \;=\; 63$

$\qquad a \;=\; \boxed{}$

5 What is the value of x or y?

a $\quad 21 \;\div\; y \;=\; 7$

$\qquad y \;=\; \boxed{}$

c $\quad 40 \;\div\; x \;=\; 8$

$\qquad x \;=\; \boxed{}$

b $\quad x \;\div\; 6 \;=\; 5$

$\qquad x \;=\; \boxed{}$

d $\quad y \;\div\; 6 \;=\; 9$

$\qquad y \;=\; \boxed{}$

6 Find the value of the letters in these mixed multiplication and division equations.

a $\quad 48 \;\div\; b \;=\; 12$

$\qquad b \;=\; \boxed{}$

e $\quad 2x \;\div\; 8 \;=\; 7$

$\qquad x \;=\; \boxed{}$

b $\qquad 8x \;=\; 48 \qquad$ ($8x$ means the same as 8 x x)

$\qquad x \;=\; \boxed{}$

c $\qquad 12a \;=\; 60$

$\qquad a \;=\; \boxed{}$

Clue: Whatever you do to one side you must do to the other.

Example: $4x = 16$

To find the value of x, divide both sides by 4.

So $x = 4$

d $\quad y \;\div\; 9 \;=\; 4$

$\qquad y \;=\; \boxed{}$

7 Find the value of the letters in these equations. Take care - they are trickier and the operations are all mixed.

a $54 + x = 96$

$x =$ ☐

b $b \div 8 = 12$

$b =$ ☐

c $y - 57 = 32$

$y =$ ☐

d $6a = 72$

$a =$ ☐

e $252 - b = 41$

$b =$ ☐

f $7y = 84$

$y =$ ☐

g $160 \div a = 8$

$a =$ ☐

h $5x + 123 = 478$

$x =$ ☐

i $3y - 27 = 66$

$y =$ ☐

j $6 \times 2a = 120$

$a =$ ☐

k $6b \div 18 = 3$

$b =$ ☐

l $39 + 9x = 84$

$x =$ ☐

Name .... Class .... Date ....

Solve equations where letters represent missing numbers.

Remember, in algebra we can use letters to represent missing numbers.

1 Find the value of x or y in these equations. Put your answers in the boxes.

a $\quad x \quad + \quad 52 \quad = \quad 197$

$\qquad x \quad = \quad \boxed{}$

b $\quad 74 \quad + \quad y \quad = \quad 188$

$\qquad y \quad = \quad \boxed{}$

c $\quad y \quad + \quad 126 \quad = \quad 387$

$\qquad y \quad = \quad \boxed{}$

d $\quad 648 \quad + \quad x \quad = \quad 828$

$\qquad x \quad = \quad \boxed{}$

2 Now find the value of a or b in these.

a $\quad 178 \quad - \quad x \quad = \quad 56$

$\qquad x \quad = \quad \boxed{}$

b $\quad y \quad - \quad 412 \quad = \quad 76$

$\qquad y \quad = \quad \boxed{}$

c $\quad 743 \quad - \quad y \quad = \quad 471$

$\qquad y \quad = \quad \boxed{}$

d $\quad x \quad - \quad 354 \quad = \quad 237$

$\qquad x \quad = \quad \boxed{}$

3 Find the value of the letters in these mixed addition and subtraction equations.

a $\quad 127 \quad + \quad x \quad = \quad 436$

$\qquad x \quad = \quad \boxed{}$

b $\quad y \quad - \quad 284 \quad = \quad 173$

$\qquad y \quad = \quad \boxed{}$

c $\quad 835 \quad - \quad b \quad = \quad 628$

$\qquad b \quad = \quad \boxed{}$

d $\quad y \quad + \quad 382 \quad = \quad 947$

$\qquad y \quad = \quad \boxed{}$

e $\quad a \quad - \quad 676 \quad = \quad 283$

$\qquad a \quad = \quad \boxed{}$

f $\quad x \quad + \quad 423 \quad = \quad 1638$

$\qquad x \quad = \quad \boxed{}$

Name .... Class .... Date ....

4 Now try to find a or b in these equations.

a a x   12   =   144   **c** b x   8   =   112

          a =   [   ] b =   [   ]

b   11   x b =   143   **d**   5   x a =   950

          b =   [   ] a =   [   ]

5 What is the value of x or y?

a   238   ÷ y =   7   **c**   896   ÷ x =   8

          y =   [   ] x =   [   ]

b x ÷   9   =   74   **d** y ÷   13   =   27

          x =   [   ] y =   [   ]

6 Find the value of the letters in these mixed multiplication and division equations.

a y ÷   12   =   79   **e** $3a$ ÷   27   =   14

          y =   [   ] a =   [   ]

b $9a$ =   279   ($9a$ means the same as 9 x a)

          a =   [   ]

c   315   ÷ b =   15

          b =   [   ]

d $12x$ =   672

          x =   [   ]

> **Clue:** Whatever you do to one side you must do to the other.
>
> **Example:** $4x = 16$
>
> To find the value of x, divide both sides by 4.
>
> **So** $x = 4$

Name .... Class .... Date ....

7 Find the value of the letters in these equations. Take care - they are trickier and the operations are all mixed.

a $436 + x = 843$

$x = \boxed{}$

g $7668 \div 2a = 9$

$a = \boxed{}$

b $b \div 52 = 48$

$b = \boxed{}$

h $4x + 637 = 2329$

$x = \boxed{}$

c $y - 158 = 674$

$y = \boxed{}$

i $2y - 749 = 873$

$y = \boxed{}$

d $15a = 840$

$a = \boxed{}$

j $594 \div 6a = 33$

$a = \boxed{}$

e $426 - b = 178$

$b = \boxed{}$

k $2b \div 46 = 137$

$b = \boxed{}$

f $21y = 882$

$y = \boxed{}$

l $326.8 + 7x = 482.9$

$x = \boxed{}$

(ch) ALGEBRA

Solve equations where letters represent missing lengths.

1 Find the value of x or y in these equations. Put your answers in the boxes.

a $12\text{cm} + x = 19\text{cm}$

$x = $ [_____] cm

b $y - 4\text{m} = 16\text{m}$

$y = $ [_____] m

c $38\text{mm} - y = 12\text{mm}$

$y = $ [_____] mm

d $b + 12\text{km} = 84\text{km}$

$b = $ [_____] km

2 Now find the missing lengths in these multiplication or division equations.

a $8\text{km} \times y = 48\text{km}$

$y = $ [_____] km

b $48\text{cm} \div a = 4\text{cm}$

$a = $ [_____] cm

c $x \div 5\text{m} = 15\text{m}$

$x = $ [_____] m

d $b \times 7\text{mm} = 84\text{mm}$

$b = $ [_____] mm

3 What are the missing lengths?

a $82\text{m} - x = 21\text{m}$

$x = $ [_____] m

b $24\text{km} + a = 77\text{km}$

$a = $ [_____] km

c $y \times 12\text{cm} = 96\text{cm}$

$y = $ [_____] cm

d $2x \div 8\text{m} = 14\text{m}$

$x = $ [_____] m

e $4b + 13\text{mm} = 33\text{mm}$

$b = $ [_____] mm

f $3y - 7\text{m} = 32\text{m}$

$y = $ [_____] m

ALGEBRA

Solve equations where letters represent missing lengths.

1 What are the missing lengths? Put your answers in the boxes.

a $72\text{cm} + x = 196\text{cm}$

$x =$ [___ cm]

c $378\text{mm} - y = 158\text{mm}$

$y =$ [___ mm]

b $a - 72\text{cm} = 123\text{m}$

$a =$ [___ m]

d $b + 426\text{km} = 513\text{km}$

$b =$ [___ km]

2 Now find the missing lengths in these multiplication or division equations.

a $7\text{km} \times y = 735\text{km}$

$y =$ [___ km]

c $338\text{m} \div x = 26\text{m}$

$x =$ [___ m]

b $a \div 8\text{cm} = 54\text{cm}$

$a =$ [___ cm]

d $b \times 46\text{mm} = 736\text{mm}$

$b =$ [___ mm]

3 What are the missing lengths?

a $552\text{m} - x = 327\text{m}$

$x =$ [___ m]

d $4b \div 72\text{m} = 432\text{m}$

$b =$ [___ m]

b $413\text{km} + a = 1623\text{km}$

$a =$ [___ km]

e $2x + 14.34\text{cm} = 28.54\text{cm}$

$x =$ [___ cm]

c $y \times 17\text{cm} = 595\text{cm}$

$y =$ [___ cm]

f $5y \times 3.2\text{m} = 112\text{m}$

$y =$ [___ m]

(ch) ALGEBRA

Name .... Class .... Date ....

Solve equations where letters represent missing lengths in shapes.

1 Look at the triangle below. Use the equation shown to find the length of side A.

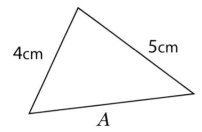

$$4\text{cm} + 5\text{cm} + A = 15\text{cm}$$

$$A = \boxed{} \text{cm}$$

2 What is the length of the missing side in April's garden?

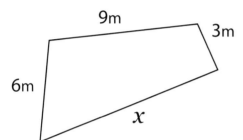

$$28\text{m} = x + 9\text{m} + 3\text{m} + 6\text{m}$$

$$\boxed{} \text{m} = x$$

3 y centimetres represents the length of each side of a regular octagon. The perimeter of the octagon is 24mm.

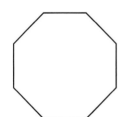

so $8y = 24\text{mm}$ ($8y$ means the same as $8 \times y$)

$$y = \boxed{} \text{mm}$$

4 Look at the track. Use the equation to find the length of x.

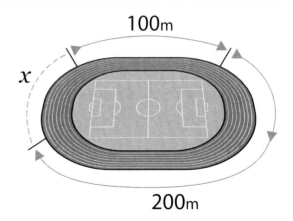
100m

200m

$$400\text{m} = 100\text{m} + x + 200\text{m}$$

$$\boxed{} \text{m} = x$$

ALGEBRA

Name .... Class .... Date ....

Solve equations where letters represent missing lengths in shapes.

1 Look at the shape below. Use the equation shown to find the length of side B.

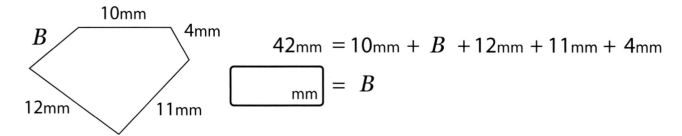

$42\text{mm} = 10\text{mm} + B + 12\text{mm} + 11\text{mm} + 4\text{mm}$

$\boxed{}\,_{\text{mm}} = B$

2 What is the length of the missing side in the model playground below? Look carefully at the measurements.

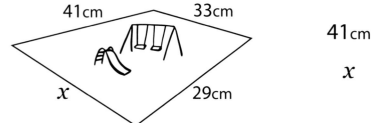

$41\text{cm} + 33\text{cm} + 29\text{cm} + x = 1.5\text{m}$

$x = \boxed{}\,_{\text{cm}}$

3 The perimeter of a regular nonagon-shaped lawn is 40.5 metres. The length of each side of the lawn is represented by y metres.

so $9y = 40.5\text{m}$ (9y means the same as 9 x y)

$y = \boxed{}\,_{\text{m}}$

4 Look at the lake. Use the equation to find the length of x.

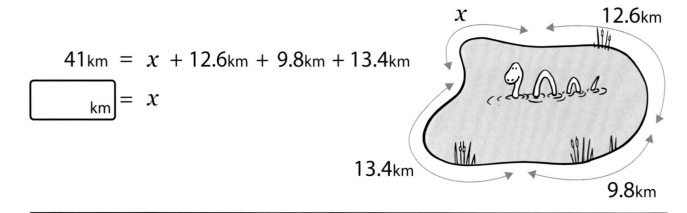

$41\text{km} = x + 12.6\text{km} + 9.8\text{km} + 13.4\text{km}$

$\boxed{}\,_{\text{km}} = x$

(ch) ALGEBRA

Solve equations where letters represent missing co-ordinates..

1 Look at the co-ordinate grid and use it to answer the questions below.

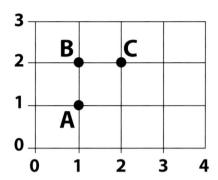

---

a The co-ordinates of **A** are (1,1). What are the co-ordinates of **C**?

C = (.......... , ...)

---

b Shape **ABCD** is a square. What are the co-ordinates of **D**?

D = (.......... , ...)

---

c **F** is exactly the same distance from **C** as is **B**. What is the missing co-ordinate?

F = (.......... , 2)

---

d Shape **ABFG** is a rectangle. What are the co-ordinates of **G**?

G = (.......... , ...)

---

e If lines were drawn on the grid to make shape **ABG**, what would the shape be?

....

Continued overleaf ALGEBRA

2 Look at the co-ordinate grid and use it to answer the questions below.

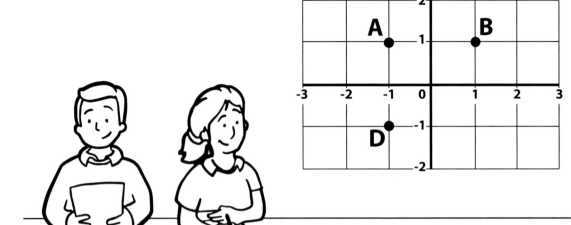

a What are the co-ordinates of **A** and **B**?

A = (.......... , ...)

B = (.......... , ...)

- -

b Shape **ABCD** is a square. What are the co-ordinates of **C**?

C = (.......... , ...)

- -

c The triangle **ADF** is drawn on the grid. What could the missing co-ordinates be?

F = (2 , ...)

or F = (2 , ...)

or F = (2 , ...)

or F = (2 , ...)

or F = (2 , ...)

ALGEBRA

Name .... Class .... Date ....

Solve equations where letters represent missing co-ordinates..

1 Look at the co-ordinate grid and use it to answer the questions below.

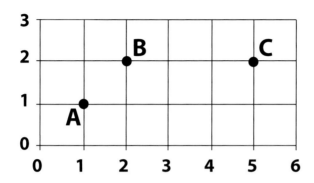

a What are the co-ordinates of **A** , **B** and **C**?

A = (.......... , ...)   **B** = (.......... , ...)   **C** = (.......... , ...)

- -

b **ABCD** is a parallelogram. What are the
co-ordinates of **D**?

D = (.......... , ...)

- -

c Line **BF** is perpendicular to the x-axis.
What is the missing co-ordinate for point **F**?

F = (.......... , 0)

- -

d A line parallel to **AB** is drawn on the grid.
The co-ordinates of one end of the line are
(2, 0). What could the co-ordinates of the
other end be (point **G**)?

G = (.......... , ...)

or **G** = (.......... , ...)

or **G** = (.......... , ...)

 Continued overleaf **ch** **ALGEBRA**

Name .... Class .... Date ....

2 Look at the co-ordinate grid and use it to answer the questions below.

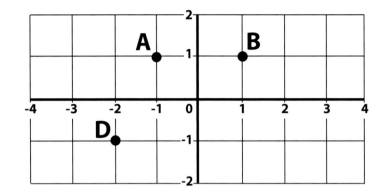

a What are the co-ordinates of **A** , **B** and **D**?

A = (.......... , ...)   **B** = (.......... , ...)   **D** = (.......... , ...)

- -

b Shape **ABCD** is a trapezium. What are the
co-ordinates of **C** if it is the same distance
from the y axis as **D**?

C = (.......... , ...)

- -

c The rectangle **ABEF** is drawn on the grid. Side **BE** is parallel to the y-axis.
What could the co-ordinates of **E** and **F** be?

E = (.......... , ...)   *and*   **F** = (.......... , ...)

or **E** = (.......... , ...)   *and*   **F** = (.......... , ...)

or **E** = (.......... , ...)   *and*   **F** = (.......... , ...)

or **E** = (.......... , ...)   *and*   **F** = (.......... , ...)

(ch) ALGEBRA

Name .... Class .... Date ....

Solve equations where letters represent missing angles.

1 Use the equation to find the size of angle a.

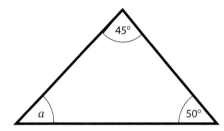

$$45° \ + \ 50° \ + \ a \ = \ 180°$$

$$a \ = \ \boxed{\qquad}°$$

---

2 Find angle x in this shape.

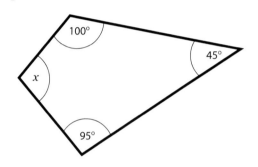

$$95° \ + \ 45° \ + \ 100° \ + \ x \ = \ 360°$$

$$x \ = \ \boxed{\qquad}°$$

---

3 Find angle b.

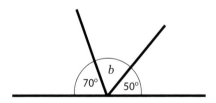

$$70° \ + \ 50° \ + \ b \ = \ 180°$$

$$b \ = \ \boxed{\qquad}°$$

---

4 Find angle y.

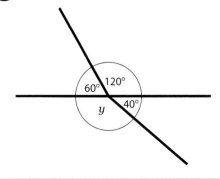

$$120° + \ 40° \ + \ y \ + \ 60° = \ 360°$$

$$y \ = \ \boxed{\qquad}°$$

ALGEBRA

Name .... Class .... Date ....

Solve equations where letters represent missing angles.

1 Use the equation to find the size of angle a.

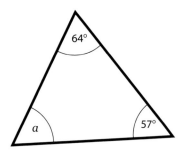

$$64° + 57° + a = 180°$$
$$a = \boxed{}°$$

---

2 Find x in this shape.

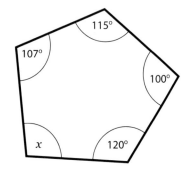

$$120° + 100° + 115° + 107° + x = 540°$$
$$x = \boxed{}°$$

---

3 Angle B, and angle D are equivalent. What is angle B?

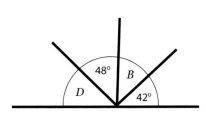

$$D = B$$
$$D + 48° + 42° + B = 180°$$
$$B = \boxed{}°$$

---

4 How many degrees is angle y?

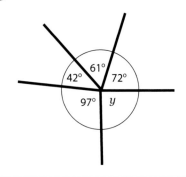

$$61° + 72° + y + 97° + 42° = 360°$$
$$y = \boxed{}°$$

(ch) ALGEBRA

Name .... Class .... Date ....

Use simple formulae.

1 Use the formula below to find the area of the following rectangles.

$$A \text{ (Area)} = l \text{ (length)} \times h \text{ (height)}$$

a $l = 3\text{cm}$
$w = 43\text{cm}$
$A = \boxed{ \text{cm}^2}$

b $l = 6\text{m}$
$w = 9\text{m}$
$A = \boxed{ \text{m}^2}$

c $l = 10\text{mm}$
$w = 46\text{mm}$
$A = \boxed{ \text{mm}^2}$

---

2 Now use the formula to find the missing width or length.

a $l = 8\text{cm}$
$w = \boxed{ \text{cm}}$
$A = 24\text{cm}^2$

b $l = \boxed{ \text{m}}$
$w = 9\text{m}$
$A = 63\text{m}^2$

c $l = 15\text{cm}$
$w = \boxed{ \text{cm}}$
$A = 30\text{cm}^2$

---

3 Use the formula below to find the missing values in the table.

$$V \text{ (Volume)} = (l) \text{ length} \times (w) \text{ width} \times (h) \text{ height}$$

	l	w	h	V
Cuboid A	3cm	2cm	1cm	$\boxed{ \text{cm}^3}$
Cuboid B	5m	$\boxed{ \text{m}}$	4m	20m^3
Cuboid C	3mm	2mm	$\boxed{ \text{mm}}$	36mm^3
Cuboid D	$\boxed{ \text{cm}}$	2cm	10cm	60cm^3

Name .... Class .... Date ....

---

4 The formula for the area of a parallelogram is $A = b$ (base) x h. Use this to complete the table below.

	A	b	h
Parallelogram A	16cm^2	4cm	☐ cm
Parallelogram B	☐ m^2	8m	6m
Parallelogram C	54mm^2	☐ mm	6mm

---

5 Use the formula below to find the missing values in the table below.

$$A = \frac{b \times h}{2} \quad \text{or} \quad \frac{1}{2} \times b \times h$$

	A	b	h
Triangle A	☐ cm^2	7cm	2cm
Triangle B	6cm^2	3cm	☐ cm
Triangle C	10m^2	☐ m	5m

---

6 In the formula $D = 2r$, D represents the diameter of a circle and r represents the radius. Use the formula to find the missing values.

	D	r
Circle A	☐ cm	4cm
Circle B	12m	☐ m
Circle C	24cm	☐ cm
Circle D	☐ m	13m

ANGLE FACT BOX

Fact	Equation
The sum of the interior angles of a triangle is $180°$	$a + b + c = 180°$
The sum of the interior angles of a quadrilateral is $360°$	$a + b + c + d = 360°$
The sum of the interior angles of a pentagon is $540°$	$a + b + c + d + e = 540°$

---

7 Use the equations shown in the fact box above to find the missing values.

Shape	Angle A	Angle B	Angle C	Angle D	Angle E
Triangle	$64°$	$72°$	☐$°$		
Quadrilateral	$75°$	$95°$	$105°$	☐$°$	
Square	☐$°$	☐$°$	☐$°$	☐$°$	
Pentagon	$110°$	$100°$	$90°$	☐$°$	$140°$
Regular Pentagon	☐$°$	$108°$	☐$°$	☐$°$	☐$°$

Name .... Class .... Date ....

Use simple formulae.

1 Use the formula below to find the area of the following rectangles.

$$A \text{ (Area)} = l \text{ (length)} \times h \text{ (height)}$$

a l = 13cm

w = 8cm

A = [___] cm^2

b l = 27m

w = 9m

A = [___] m^2

c l = 23mm

w = 13mm

A = [___] mm^2

---

2 Now use the formula to find the missing width or length.

a l = 14cm

w = [___] cm

A = 98cm^2

b l = [___] m

w = 16m

A = 672m^2

c l = 24cm

w = [___] cm

A = 504cm^2

---

3 Use the formula below to find the missing values in the table.

$$V \text{ (Volume)} = (l) \text{ length} \times (w) \text{ width} \times (h) \text{ height}$$

	l	w	h	V
Cuboid A	4cm	[___] cm	7cm	56cm^3
Cuboid B	8m	6m	9m	[___] m^3
Cuboid C	4.5mm	2mm	[___] mm	108mm^3
Cuboid D	[___] cm	4cm	7cm	252cm^3

4 The formula for the area of a parallelogram is $A = b$ (base) x h. Use this to complete the table below.

	A	b	h
Parallelogram A	156cm^2	12cm	cm
Parallelogram B	m^2	15m	16m
Parallelogram C	374mm^2	22mm	mm

5 Use the formula below to find the missing values in the table below.

$$A = \frac{b \times h}{2} \text{ or } \frac{1}{2} \times b \times h$$

	A	b	h
Triangle A	cm^2	8cm	29cm
Triangle B	300cm^2	5cm	cm
Triangle C	496m^2	m	16m

6 In the formula $D = 2r$, D represents the diameter of a circle and r represents the radius. Use the formula to find the missing values.

	D	r
Circle A	m	57m
Circle B	1240cm	cm
Circle C	265m	m
Circle D	cm	7.25cm

Continued overleaf (ch) ALGEBRA

ANGLE FACT BOX

Fact	Equation
The sum of the interior angles of a triangle is 180°	$a + b + c = 180°$
The sum of the interior angles of a quadrilateral is 360°	$a + b + c + d = 360°$
The sum of the interior angles of a pentagon is 540°	$a + b + c + d + e = 540°$

7 Use the equations shown in the fact box above to find the missing values.

Shape	Angle A	Angle B	Angle C	Angle D	Angle E
Triangle	62.5°	78°	☐°		
Equilateral Triangle	☐°	☐°	☐°		
Square	☐°	☐°	☐°	☐°	
Pentagon	116°	104.5°	87.5°	☐°	141°
Regular Pentagon	☐°	☐°	☐°	☐°	☐°

Continued overleaf (ch) ALGEBRA

Name .... Class .... Date ....

Look at the formula. n represents the number of sides.

$$(n - 2) \times 180° = \text{sum of the interior angles of any polygon}$$

8 Use this formula to complete the missing values in the polygons below.

	Number of sides (name of shape)	Sum of interior angles	Each interior angle of the regular polygon
a	9 (nonagon)	[ ]°	[ ]°
b	10 (decagon)	[ ]°	[ ]°
c	[ ]	1080°	[ ]°

d Now use the formula to find the missing angle in the irregular hexagon below.

Angle A = 115°   Angle B = 127°

Angle C = 120°   Angle D = 131°

Angle E = 112°   Angle F = [   ]°

e Write down a possible combination for the angles in a pentagon (5 sides).

Well done!

Generate and describe linear number sequences.

1 For each number sequence below, find the rule to describe the pattern and write the missing numbers.

EXAMPLE

4,   8,   12,   16, | 20 | | 24 | | 28 |

✏️ **add 4**

- -

a 9,   14,   19,   24,   [ ]   [ ]   [ ]

✏️

b 45,   54,   [ ]   72,   [ ]   90,   [ ]

✏️

c 78   67,   [ ]   45,   34   [ ]   [ ]

✏️

d 5,   2,   -1,   -4,   [ ]   [ ]   [ ]

✏️

e 125,   [ ]   75,   50,   [ ]   0,   [ ]

✏️

Name... Class........................ Date........................

2 Now try these. **Clue** - the pattern isn't just adding or subtracting.

a 2, 4, 8, ⬚ 32, ⬚ ⬚

b 1, 4, 16, 64, ⬚ ⬚ ⬚

c 729, 243, ⬚ 27, 9, ⬚ ⬚

d 320, 160, 80, ⬚ 20, ⬚ ⬚

e 0.04, 0.2, 1, 5, ⬚ ⬚ ⬚

f 1.25, 2.5, 5, 10, ⬚ ⬚ ⬚

Continued overleaf ALGEBRA

Name .... Class .... Date ....

3 For these, you will need to find the pattern in the difference between the numbers.

EXAMPLE

2,   3,   5,   8,   [12]   [17]   [23]

✏️ (increase) | decrease difference by (1) | 2 and (add) | subtract

- -

a   8,   10,   13,   17,   [ ]   [ ]   [ ]

✏️ increase | decrease difference by 1 | 2 and add | subtract

b   25,   24,   22,   19,   [ ]   [ ]   [ ]

✏️ increase | decrease difference by 1 | 2 and add | subtract

c   2,   12,   21,   29,   [ ]   [ ]   [ ]

✏️ increase | decrease difference by 1 | 2 and add | subtract

4 Now have a go at writing the full rule and finding the missing numbers.

a   1,   2,   5,   10,   [ ]   [ ]   [ ]

✏️ increase | decrease difference by 1 | 2 and add | subtract

b   50,   40,   31,   23,   [ ]   [ ]   [ ]

✏️ increase | decrease difference by 1 | 2 and add | subtract

Name .... Class .... Date ....

5 Describe how these numbers are special and complete the sequence.

a   1,   4,   9,   [ ]   25,   [ ]   [ ]

✏ ....

b   2,   3,   7,   [ ]   13,   17,   [ ]   [ ]

✏ ....

---

6 These are tricky! There are two rules.

EXAMPLE

$+3$ -2 $+3$ -2 $+3$ -2 $+3$

2,   5,   3,   6,   4,   [7]   [5]   [8]

✏ **add 3, subtract 2** ....

- -

a   4,   5,   3,   4,   2,   [ ] [ ] [ ]

✏ ....

b   3,   8,   12,   17,   21,   [ ] [ ] [ ]

✏ ....

c   25,   20,   22,   17,   19,   [ ] [ ] [ ]

✏ ....

Name .... Class .... Date ....

Generate and describe linear number sequences.

1 For each number sequence below, find the rule to describe the pattern and write the missing numbers.

EXAMPLE

8,   14,   20,   26, | 32 | | 38 | | 44 |

✏️ **add 6**
....

- -

a) 61,   68, | | 82, | | | | | |

✏️ ....

b) 128, | | 106,   95, | | | | | |

✏️ ....

c) 17,   20, | | | | 29, | | 35, | |

✏️ ....

d) | | -3,   0, | | | | 9,   12, | |

✏️ ....

e) 65, | | | | 35, | | | | 5, | |

✏️ ....

2 Now try these. **Clue** - the pattern isn't just adding or subtracting.

a 4, ☐ ☐ 32, ☐ 128, ☐

b 8, ☐ 72, 216, ☐ 1944 ☐

c 176, ☐ 44, ☐ 11, 5.5, ☐

d 0.25, ☐ 1, ☐ 4, ☐ ☐

e 48, ☐ 12, ☐ ☐ 1.5 ☐

f ☐ ☐ 2, ☐ 0.5, 0.25, ☐

69 *Continued overleaf* **ch** **ALGEBRA**

3 For these, you will need to find the pattern in the difference between the numbers.

EXAMPLE

  1   2   3   4   5   6

24,   25,   27,   30, | 34 | | 39 | | 45 |

✏️ (increase) | decrease difference by (1) | 2 | 3 and (add) | subtract
....

- -

a 56,   [ ]   53,   50,   [ ] [ ]   35,   [ ]

✏️ increase | decrease difference by 1 | 2 | 3 and add | subtract
....

b 5,   15,   24,   [ ] [ ] [ ] [ ] [ ]

✏️ increase | decrease difference by 1 | 2 | 3 and add | subtract
....

4 Now have a go at writing the full rule and finding the missing numbers.

a 11,   10,   [ ]   5,   [ ]   -4,   [ ] [ ]

✏️ increase | decrease difference by 1 | 2 | 3 and add | subtract
....

b [ ]   40,   33,   27,   22,   [ ] [ ] [ ]

✏️ increase | decrease difference by 1 | 2 | 3 and add | subtract
....

c [ ]   3,   1,   -2,   [ ]   -11,   [ ] [ ]

✏️ increase | decrease difference by 1 | 2 | 3 and add | subtract
....

70 *Continued overleaf* (ch) ALGEBRA

5 Describe how these numbers are special and complete the sequence.

a [ ] 4, [ ] 16, [ ] [ ] 49, [ ]

✎ ....

b | 37 | | | | 19 | | | |
   |----|----|----|----|----|----|----|----|

✎ ....

---

6 These are tricky! There are two rules.

EXAMPLE

+4   -5   +4   -5   +4   -5

16	20	15	19	14	18	13

✎ **add 4, subtract 5** ....

- -

a 13,   17,   9,   [ ] 5,   [ ] [ ] 5,   [ ]

✎ ....

b 8,   16,   14,   28,   [ ] [ ] [ ] [ ] [ ]

✎ ....

c 5,   15,   17,   [ ] 53,   [ ] [ ] [ ] [ ]

✎ ....

Name... Class........................ Date........................

Use a formula to generate a number in a number sequence.

1 The formula for the sequence below is:

$$(n \times 3) + 1 \qquad (n = 1^{st}, 2^{nd}, 3^{rd}, 4^{th} \text{ number etc })$$

4	**7**	**10**	**13**	**16**
1st	2nd	3rd	4th	5th

Use the formula to find the missing numbers in the sequence.

EXAMPLE

8th → $(8 \times 3) + 1 = 25$

a 5th → [ ]　　　**d** 10th → [ ]

b 2nd → [ ]　　　**e** 1st → [ ]

c 9th → [ ]　　　**f** 12th → [ ]

- -

2 Now use the formula **($n \times 5$) + 3** to fnd the missing numbers.

a 1st → [ ]　　　**d** 8th → [ ]

b 3rd → [ ]　　　**e** 10th → [ ]

c 5th → [ ]　　　**f** 14th → [ ]

You're doing a great job!

ALGEBRA

Name .... Class .... Date ....

Use a formula to generate a number in a number sequence.

1 The formula for the sequence **10**, **17**, **24**, **31** is $(n \times 7) + 3$, where n represents the 1st, 2nd, 3rd number etc.

Use the formula to find the missing numbers in the sequence.

EXAMPLE

8th ➜ $(8 \times 7) + 3 = 59$

(a) 12th ➜ [ ]　(b) 15th ➜ [ ]

- -

2 Now use the formula $n^2 \times 2$

(a) 1st ➜ [ ]　(c) 10th ➜ [ ]

(b) 7th ➜ [ ]　(d) 12th ➜ [ ]

- -

3 Have a go at using the formula $n^3 - 8$

(a) 4th ➜ [ ]　(c) 1st ➜ [ ]

(b) 5th ➜ [ ]　(d) 2nd ➜ [ ]

- -

4 Write the first 5 numbers in the sequence using the formula $(n \times 9) - 20$

[ ]　[ ]　[ ]

[ ]　[ ]

(ch) ALGEBRA

Name .... Class .... Date ....

Express missing number problems algebraically.

---

1 Complete the following.

a Jaz had 16 precious coins. He gave x coins to Afzal. He had 7 left. **Circle** the equation which would show how many precious coins Jaz gave to Afzal.

$$16 \div x = 7 \qquad\qquad 16 - x = 7 \qquad\qquad 16 + 7 = x$$

---

b David bought 8 packets of biscuits. Each packet had the same number of biscuits. There were 208 biscuits altogether. **Circle** the equation which could be used to work out how many biscuits were in each packet (y).

$$y = 8 \times 208 \qquad\qquad y = 208 - 8 \qquad\qquad y = 208 \div 8$$

---

c There are 18 tennis balls in the bag. 7 are green and x are orange. **Circle** the equation which could be used to find x.

$$x - 18 = 7$$

$$18 = 7 + x$$

$$x = 7 - 18$$

---

d Molly has £4. She spends £y at the shop. She has £1.50 left. **Circle** the equation which could be used to find y.

$$£y = £4 - £1.50 \qquad\qquad £1.50 + £4 = £y \qquad\qquad £4 = £1.50 - £y$$

Continued overleaf ALGEBRA

Name .... Class .... Date ....

2 Write down a suitable equation you could use to solve the problems. Then solve them.

a A toy shop had 8 toy garages. Each garage had 4 toy cars. How many cars were there altogether (*a*)?

$a = \boxed{}\ \boxed{}\ \boxed{}$ $\qquad$ $a = \boxed{}$

---

b A wall had 48 bricks. There were 23 old bricks and the rest were new. How many new bricks (*b*) were used in the wall?

$b = \boxed{}\ \boxed{}\ \boxed{}$ $\qquad$ $b = \boxed{}$

---

c The Crafty Cake Shop sells 4 birthday cakes on Monday. On Tuesday, it sells 5 times as many. How many birthday cakes did the shop sell on Tuesday (*x*)?

$x = \boxed{}\ \boxed{}\ \boxed{}$ $\qquad$ $x = \boxed{}$

---

d Holly buys 27 sweets. Harry buys ⅓ as many. How many sweets does Harry buy (*s*)?

$s = \boxed{}\ \boxed{}\ \boxed{}$ $\qquad$ $s = \boxed{}$

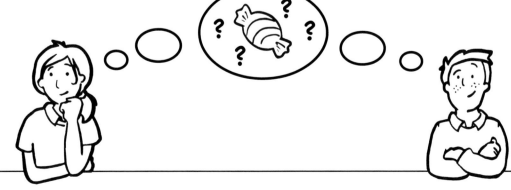

ALGEBRA

Name .... Class .... Date ....

Express missing number problems algebraically.

1 Complete the following.

a A bag contains x nails. Charlie used 56 to build a cabinet. There were 127 nails left in the bag. **Circle** the equation which could be used to work out how many nails were in the bag to start with.

$x = 127 - 56$

$x = 56 + 127$

$x = 56 \times 127$

---

b Gerry put the buns he had made into 14 tins. He had made 252 buns. **Circle** the equation which could be used to work out how many buns were in each tin (y).

$14 + 252 = y$ $\qquad$ $252 \times 14 = y$ $\qquad$ $252 \div 14 = y$

---

c Four friends went to a football match. They each paid £4.50 for a drink and a burger. **Circle** the equation which could be used to work out the total bill (b).

£4.50 $+ 4 = b$ $\qquad$ £4.50 $\div 4 = b$ $\qquad$ £4.50 $\times 4 = b$

---

d On Saturday 1364 people visited the Green Garden Centre. On Sunday 956 people visited. Circle the equation which could be used to work out how many people visited the garden centre altogether on Saturday and Sunday (a).

$1364 + 956 = a$ $\qquad$ $1364 - 956 = a$ $\qquad$ $1364 \div a = 956$

---

2 Write down a suitable equation you could use to solve the problems. Then solve them.

a 1486 people attend the Laid Back Leisure Centre over one weekend. 652 are female. How many are male (*a*)?

a = [ ] [ ] [ ] a = [ ]

b There were three times as many tiles in the large pool as in the small pool. The small pool had 876 tiles. How many tiles were in the large pool (*b*)?

b = [ ] [ ] [ ] b = [ ]

c Mrs Shah shared £24.45 evenly between her three children. How much did they each get (*x*)?

x = [£] [ ] [ ]

x = [£]

d Megan spent £2.64 on a necklace, £1.25 on some sweets and £15.72 on a present for her mum. How much did she spend altogether (*y*)? Write the full equation on the line.

....

y = [£]

(ch) **ALGEBRA**

Name .... Class .... Date ....

Solve equivalent expressions, where a letter represents one unknown number.

1 Find the missing numbers in the following.

a $5 + x = 4 + 11$

$x =$ ⬚

b $38 - 9 = y - 3$

$y =$ ⬚

c $6a = 3 \times 12$

$a =$ ⬚

d $40 \div b = 24 \div 3$

$b =$ ⬚

e $4y = 16 \times 2$

$y =$ ⬚

f $4 + 23 = a + 21$

$a =$ ⬚

g $x \div 4 = 70 \div 7$

$x =$ ⬚

h $63 - b = 58 - 17$

$b =$ ⬚

2 Now try these.

a $15 + x = 24 - 6$

$x =$ ⬚

b $8 \times 7 = 112 \div y$

$y =$ ⬚

c $59 - x = 31 + 23$

$x =$ ⬚

d $72 \div 2 = 9b$

$b =$ ⬚

e $98 - 49 = y + 3$

$y =$ ⬚

f $a \div 3 = 6 \times 4$

$a =$ ⬚

g $137 - 14 = x + 112$

$x =$ ⬚

h $1.5b = 45 \div 3$

$b =$ ⬚

Name .... Class .... Date ....

3 Have a go at these - look carefully at the operations.

a $14 + y = 3 \times 6$

$y = \boxed{}$

e $8x = 49 + 39$

$x = \boxed{}$

b $27 \div 9 = 52 - x$

$x = \boxed{}$

f $84 - a = 17 \times 3$

$a = \boxed{}$

c $8b = 33 - 9$

$b = \boxed{}$

g $288 \div y = 181 - 37$

$y = \boxed{}$

d $15 + 25 = a - 2$

$a = \boxed{}$

h $312 \div 3 = 67 + b$

$b = \boxed{}$

4 Now try these. Think carefully - they are tricky!

a $122 + 22 = 4y = 217 - 73$

$y = \boxed{}$

b $156 - a = 243 \div 3 = 3 \times 27$

$a = \boxed{}$

c $96 \div 12 = 0.5 \times 16 = 3.2 + x$

$x = \boxed{}$

d $206.5 + 0.5 = b \div 3 = 571 - 364$

$b = \boxed{}$

ALGEBRA

Solve equivalent expressions, where a letter represents one unknown number.

1 Find the missing numbers in the following.

a $52 + x = 37 + 56$

$x =$ [ ]

b $169 - 43 = y - 42$

$y =$ [ ]

c $18a = 12 \times 9$

$a =$ [ ]

d $182 \div b = 156 \div 12$

$b =$ [ ]

e $91y = 26 \times 7$

$y =$ [ ]

f $178 + 105 = a + 94$

$a =$ [ ]

g $416 \div x = 128 \div 16$

$x =$ [ ]

h $736 - b = 437 - 188$

$b =$ [ ]

2 Now try these.

a $2917 + x = 4613 - 1212$

$x =$ [ ]

b $56 \times 73 = y \div 7$

$y =$ [ ]

c $9210 - x = 4164 + 2857$

$x =$ [ ]

d $1125 \div 25 = 18b$

$b =$ [ ]

e $9427 - 649 = y + 689$

$y =$ [ ]

f $a \div 23 = 3.5 \times 16$

$a =$ [ ]

g $9473 - 1849 = x + 3476$

$x =$ [ ]

h $2.5b = 520.8 \div 8.4$

$b =$ [ ]

Continued overleaf **ch** ALGEBRA

Name .... Class .... Date ....

3 Have a go at these - look carefully at the operations.

a $67 + y = 12 \times 13$

$y =$ [box]

b $350 \div 14 = 113 - x$

$x =$ [box]

c $567 \div b = 426 - 363$

$b =$ [box]

d $1026 + 3728 = 7214 - a$

$a =$ [box]

e $58x = 769 + 159$

$x =$ [box]

f $6842 - a = 36 \times 53$

$a =$ [box]

g $736 \div y = 47.63 - 15.63$

$y =$ [box]

h $1512 \div 27 = 39.14 + b$

$b =$ [box]

4 Now try these. Think carefully - they are tricky!

a $362.6 + 158.4 = 1042 \div y = 1341 - 820$

$y =$ [box]

b $74.439 - a = 598 \div 23 = 3.25 \times 8$

$a =$ [box]

c $3478 \div 37 = 37.6 \times 2.5 = 8.764 + x$

$x =$ [box]

d $74.7 + 18.2 = b \div 4.2 = 1714 - 1621.1$

$b =$ [box]

Find pairs of numbers that satisfy an equation with two unknowns.

1 For each of the following, find the value of $\triangle$. **Clue** - find the value of the other shapes first.

a $\square$ = 23 − 5

$\square$ = 14 + $\triangle$

$\triangle$ = $\boxed{}$

c $\hexagon$ = 4 x 9

$\hexagon$ = $\triangle$ + $\triangle$ + $\triangle$

$\triangle$ = $\boxed{}$

b $\bigcirc$ = 32 ÷ 2

$\bigcirc$ = $\triangle$ x $\triangle$

$\triangle$ = $\boxed{}$

d $\pentagon$ = 52 − 27

$\pentagon$ = 43 − $\triangle$

$\triangle$ = $\boxed{}$

2 Solve the following. **Clue** - one letter is easy to work out, so find that first.

a x = 26 + 14

x = y + y

x = $\boxed{}$

d a = 97 − 76

a = b ÷ 3

b = $\boxed{}$

b y = 8 x 8

y = 73 − a

a = $\boxed{}$

e 63 + 14 = y

85 − x = y

x = $\boxed{}$

c b = 99 ÷ 11

b = 6 + x

x = $\boxed{}$

f b = 13 x 4

b = 104 ÷ y

y = $\boxed{}$

 Continued overleaf ALGEBRA

3 Have a go at solving these.

a $17 + a = 21$

$b = 2a$

$b = \boxed{}$

b $56 \div x = 7$

$x = y + y$

$y = \boxed{}$

c $10b = 80$

$x = b + 31$

$x = \boxed{}$

d $y - 16 = 17$

$a + y = 41$

$a = \boxed{}$

4 Try these - they are getting even trickier!

a $27 \div a = 3$

$a + 2 = 44 \div b$

$b = \boxed{}$

b $14 - y = 9$

$20 + y = 31 - x$

$x = \boxed{}$

c $2b = 30$

$27 - b = 2y$

$y = \boxed{}$

d $x + 38 = 44$

$2x = y - 9$

$y = \boxed{}$

e $b - 15 = 35$

$4a = b - 18$

$a = \boxed{}$

f $3y = 60$

$y - 9 = x \div 2$

$x = \boxed{}$

Continued overleaf **ALGEBRA**

5 Have a go at these. You will need to think very carefully!

a

$$4x = 24 \div 2$$

$$22 + x = 30 - y$$

$$y = \boxed{}$$

d

$$b + 57 = 12 \times 5$$

$$5a = b + 17$$

$$a = \boxed{}$$

b

$$35 \div a = 3 + 4$$

$$4a = 23 - b$$

$$b = \boxed{}$$

e

$$x + 6 = 5 + 8$$

$$y \div 2 = 5x$$

$$y = \boxed{}$$

c

$$y - 14 = 3 \times 2$$

$$y \div 4 = 1 + x$$

$$x = \boxed{}$$

f

$$30 - 6 = 12a$$

$$13b = a + 128$$

$$b = \boxed{}$$

6 These are very challenging! Try them.

a

$$x + 4 = 4 + 11 + 5$$

$$x \div 2 = 2 + a$$

$$a = \boxed{}$$

c

$$a + 7 = 3 \times 9$$

$$y = a + a + 4$$

$$y = \boxed{}$$

b

$$4 + y + 3 = 8 \times 2$$

$$60 + x = 7y$$

$$x = \boxed{}$$

d

$$8b = 97 - 9$$

$$b - 2 = 12 - y$$

$$y = \boxed{}$$

ALGEBRA

Name... Class........................ Date........................

Find pairs of numbers that satisfy an equation with two unknowns.

1 For each of the following, find the value of $\triangle$. **Clue** - find the value of the other shapes first.

a $\square$ = 179 − 23

$\square$ = 119 + $\triangle$

$\triangle$ = [ ]

c $\hexagon$ = 9 × 12

$\hexagon$ = $\triangle$ + $\triangle$ + $\triangle$ + $\triangle$

$\triangle$ = [ ]

b $\bigcirc$ = 144 ÷ 4

$\bigcirc$ = $\triangle$ × $\triangle$

$\triangle$ = [ ]

d $\pentagon$ = 156 − 68

$\pentagon$ = 124 − $\triangle$

$\triangle$ = [ ]

2 Solve the following. **Clue** - one letter is easy to work out, so find that first.

a x = 127 + 43

x = y + y

y = [ ]

d a = 314 − 179

a = b ÷ 8

b = [ ]

b y = 13 × 13

y = 204 − a

a = [ ]

e 474 + 383 = y

1068 − x = y

x = [ ]

c b = 350 ÷ 14

b = x^2

x = [ ]

f b = 36 × 14

b = 1008 ÷ y

y = [ ]

Continued overleaf (ch) **ALGEBRA**

3 Have a go at solving these.

a $156 + a = 242$

$b = 7a$

$b = \boxed{}$

c $26b = 650$

$x = b + 987$

$x = \boxed{}$

b $992 \div x = 8$

$x = y + y$

$y = \boxed{}$

d $y - 176 = 243$

$a + y = 736$

$a = \boxed{}$

4 Try these - they are getting even trickier!

a $516 \div a = 12$

$a + 7 = 250 \div b$

$b = \boxed{}$

d $x + 1147 = 1489$

$13x = y - 63$

$y = \boxed{}$

b $407 - y = 392$

$y^2 = 307 - x$

$x = \boxed{}$

e $b - 843 = 1069$

$23a = b - 1843$

$a = \boxed{}$

c $18b = 306$

$173 - b = 12y$

$y = \boxed{}$

f $14y = 1022$

$y - 59 = x \div 27$

$x = \boxed{}$

5 Have a go at these. You will need to think very carefully!

a
$$12x = 420 \div 5$$
$$426 + x = 531 - y$$
$$y = \boxed{}$$

b
$$154 \div a = 4 + 7$$
$$26a = 578 - b$$
$$b = \boxed{}$$

c
$$y - 89 = 13 \times 25$$
$$y \div 9 = 32 + x$$
$$x = \boxed{}$$

d
$$b + 661 = 43 \times 18$$
$$14a = b + 69$$
$$a = \boxed{}$$

e
$$x + 159 = 77 + 108$$
$$y \div 8 = 17x$$
$$y = \boxed{}$$

f
$$1072 - 888 = 8a$$
$$25b = a + 1027$$
$$b = \boxed{}$$

6 These are very challenging! Try them.

a
$$x + 327 = 136 + 258 + 89$$
$$x \div 12 = 7.5 + a$$
$$a = \boxed{}$$

b
$$68 + y + 75 = 51 \times 4$$
$$317 + x = 9y$$
$$x = \boxed{}$$

c
$$a + 326 = 17 \times 26$$
$$y = a + a + 76$$
$$y = \boxed{}$$

d
$$18b = 1272 - 624$$
$$b - 6 = 20y$$
$$y = \boxed{}$$

(ch) ALGEBRA

Enumerate the possibilities of two variables.

1 Look at each of the equations below. Find 3 different pairs of values for a and b.

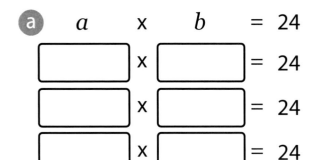

a a x b = 24

[   ] x [   ] = 24

[   ] x [   ] = 24

[   ] x [   ] = 24

e a x b = 36

[   ] x [   ] = 36

[   ] x [   ] = 36

[   ] x [   ] = 36

b a ÷ b = 8

[   ] ÷ [   ] = 8

[   ] ÷ [   ] = 8

[   ] ÷ [   ] = 8

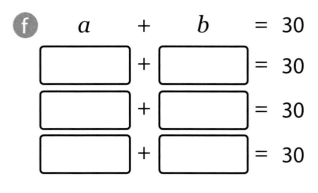

f a + b = 30

[   ] + [   ] = 30

[   ] + [   ] = 30

[   ] + [   ] = 30

c a + b = 26

[   ] + [   ] = 26

[   ] + [   ] = 26

[   ] + [   ] = 26

g a − b = 6

[   ] − [   ] = 6

[   ] − [   ] = 6

[   ] − [   ] = 6

d a − b = 4

[   ] − [   ] = 4

[   ] − [   ] = 4

[   ] − [   ] = 4

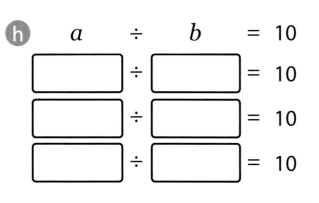

h a ÷ b = 10

[   ] ÷ [   ] = 10

[   ] ÷ [   ] = 10

[   ] ÷ [   ] = 10

Continued overleaf
ALGEBRA

Name .... Class .... Date ....

2 For these, find the different values of x and y.

a x $\times$ y $=$ 28

[   ] $\times$ [   ] $=$ 28

[   ] $\times$ [   ] $=$ 28

[   ] $\times$ [   ] $=$ 28

e x $+$ y $=$ 27

[   ] $+$ [   ] $=$ 27

[   ] $+$ [   ] $=$ 27

[   ] $+$ [   ] $=$ 27

b x $\times$ y $=$ 40

[   ] $\times$ [   ] $=$ 40

[   ] $\times$ [   ] $=$ 40

[   ] $\times$ [   ] $=$ 40

[   ] $\times$ [   ] $=$ 40

f x $-$ y $=$ 9

[   ] $-$ [   ] $=$ 9

[   ] $-$ [   ] $=$ 9

[   ] $-$ [   ] $=$ 9

[   ] $-$ [   ] $=$ 9

c x $\div$ 2 $=$ y

[   ] $\div$ 2 $=$ y

[   ] $\div$ 2 $=$ y

[   ] $\div$ 2 $=$ y

[   ] $\div$ 2 $=$ y

d x $+$ y $+$ 3 $=$ 16

[   ] $+$ [   ] $+$ 3 $=$ 16

[   ] $+$ [   ] $+$ 3 $=$ 16

[   ] $+$ [   ] $+$ 3 $=$ 16

3 Try these.

a There are 6 sweets in a bag. Some are chewy and some are soft. Use the equation $c + s = 6$ and list all the possible combinations of chewy and soft sweets.

b $xy = 12$. In the space below, list all the possible combinations of x and y (x and y are whole numbers).

c There were 30 children at a party. The number of boys (b) was greater than 9 and fewer than 16, and the number of girls was greater than 14 and fewer than 21. Use the equation $b + g = 30$ to find all the possible combinations of boys and girls.

Enumerate the possibilities of two variables.

1 Look at each of the equations below. Find 3 different pairs of values for a and b.

a

$$a \quad \times \quad b \quad = 96$$

$$\boxed{} \times \boxed{} = 96$$
$$\boxed{} \times \boxed{} = 96$$
$$\boxed{} \times \boxed{} = 96$$

e

$$a \quad \times \quad b \quad = 110$$

$$\boxed{} \times \boxed{} = 110$$
$$\boxed{} \times \boxed{} = 110$$
$$\boxed{} \times \boxed{} = 110$$

b

$$a \quad \div \quad b \quad = 12$$

$$\boxed{} \div \boxed{} = 12$$
$$\boxed{} \div \boxed{} = 12$$
$$\boxed{} \div \boxed{} = 12$$

f

$$a \quad + \quad b \quad = 155$$

$$\boxed{} + \boxed{} = 155$$
$$\boxed{} + \boxed{} = 155$$
$$\boxed{} + \boxed{} = 155$$

c

$$a \quad + \quad b \quad = 77$$

$$\boxed{} + \boxed{} = 77$$
$$\boxed{} + \boxed{} = 77$$
$$\boxed{} + \boxed{} = 77$$

g

$$a \quad - \quad b \quad = 39$$

$$\boxed{} - \boxed{} = 39$$
$$\boxed{} - \boxed{} = 39$$
$$\boxed{} - \boxed{} = 39$$

d

$$a \quad - \quad b \quad = 17$$

$$\boxed{} - \boxed{} = 17$$
$$\boxed{} - \boxed{} = 17$$
$$\boxed{} - \boxed{} = 17$$

h

$$a \quad \div \quad b \quad = 13$$

$$\boxed{} \div \boxed{} = 13$$
$$\boxed{} \div \boxed{} = 13$$
$$\boxed{} \div \boxed{} = 13$$

Continued overleaf (ch) **ALGEBRA**

Name.. Class........................ Date........................

2 For these, find the different values of x and y.

a x x y = 156

$\boxed{}$ x $\boxed{}$ = 156

$\boxed{}$ x $\boxed{}$ = 156

$\boxed{}$ x $\boxed{}$ = 156

e x + y = 253

$\boxed{}$ + $\boxed{}$ = 253

$\boxed{}$ + $\boxed{}$ = 253

$\boxed{}$ + $\boxed{}$ = 253

b x x y = 250

$\boxed{}$ x $\boxed{}$ = 250

$\boxed{}$ x $\boxed{}$ = 250

$\boxed{}$ x $\boxed{}$ = 250

$\boxed{}$ x $\boxed{}$ = 250

f x + y = 127

$\boxed{}$ + $\boxed{}$ = 127

$\boxed{}$ + $\boxed{}$ = 127

$\boxed{}$ + $\boxed{}$ = 127

$\boxed{}$ + $\boxed{}$ = 127

c x ÷ y = 25

$\boxed{}$ ÷ $\boxed{}$ = 25

$\boxed{}$ ÷ $\boxed{}$ = 25

$\boxed{}$ ÷ $\boxed{}$ = 25

$\boxed{}$ ÷ $\boxed{}$ = 25

d x + y + 57 = 94

$\boxed{}$ + $\boxed{}$ + 57 = 94

$\boxed{}$ + $\boxed{}$ + 57 = 94

$\boxed{}$ + $\boxed{}$ + 57 = 94

Continued overleaf **ch** ALGEBRA

Name .... Class .... Date ....

3 Try these.

a There were 16 biscuits in a packet. Some were plain (p) and some were chocolate (c). There were less than 8 chocolate biscuits. Use the equation $p + c = 16$ and list all the possible combinations of plain and chocolate biscuits.

b $xy = 20$. In the space below, list all the possible combinations of x and y (x and y are whole numbers).

c There were 40 children at the youth club. The number of boys (b) was greater than 19 and fewer than 26 and the number of girls (g) was greater than 14 and fewer than 21. Use the equation $g + b = 40$ to list all the possible combinations of boys and girls.

Name .... Class .... Date ....

Solve equations using the correct order of operations.

To make sure we carry out calculations in the correct order we use **BODMAS**.

B O D M A S

brackets
2 (3 + 4)
means the same
as 2 x (3 + 4)

order
(or other things
e.g. - squares)

division   **multiplication**   **addition**   **subtraction**

1 Use **BODMAS** to find the missing numbers in these equations.

a x = (8 + 2) + 3

x = ☐

b y = 15 ÷ (1 + 2)

y = ☐

c a = 4 + 6 x 8

a = ☐

d b = 30 ÷ 2 – 6

b = ☐

e y = 14 – (8 + 3)

y = ☐

f a = 7 x 4 + 3

a = ☐

g   52 – 42 ÷ 7 = x

x = ☐

h b = 27 x (16 – 15)

b = ☐

You're doing
great!

94   *Continued overleaf*   ALGEBRA

Name.. Class....................... Date.......................

2 Try these. Think carefully.

a $(3 \times 4) + 8 = x$

$x = \boxed{}$

b $y = 3 + 6 \times 5$

$y = \boxed{}$

c $a = 2 (12 + 36)$

$a = \boxed{}$

d $b = (14 + 10) \div (10 - 2)$

$b = \boxed{}$

e $2x = (2 \times 3) + 10$

$x = \boxed{}$

f $4a = 5 + (27 \div 9)$

$a = \boxed{}$

g $3y = (24 \div 3) + 7$

$y = \boxed{}$

h $3b = 3(5 + 1)$

$b = \boxed{}$

i $(17 - 6)4 = 4x$

$x = \boxed{}$

j $y = (24 - 4) \div (2 + 3)$

$y = \boxed{}$

k $4b = 4 + 24 \div 3 + 4$

$b = \boxed{}$

l $5x = 29 + 3(4 - 2)$

$x = \boxed{}$

m $4a = (27 \div 3)4 + 8$

$a = \boxed{}$

n $9y = 14 + 8(12 \div 2) - 8$

$y = \boxed{}$

Solve equations using the correct order of operations.

To make sure we carry out calculations in the correct order, we use **BODMAS**.

B O D M A S

brackets	**order**	**division**	**multiplication**	**addition**	**subtraction**

$2(3+4)$ means the same as $2 \times (3+4)$

(or other things e.g. - squares)

1 Use **BODMAS** to find the missing numbers in these equations.

a $x = (27 + 14) + 8$

$x = \boxed{}$

b $y = 48 \div (4 + 2)$

$y = \boxed{}$

c $a = 14 + 36 \times 3$

$a = \boxed{}$

d $b = 56 \div 8 \times 15$

$b = \boxed{}$

e $y = 156 - 12 \times 12$

$y = \boxed{}$

f $a = (158 + 10) \div 12$

$a = \boxed{}$

g $372 - 162 \div 9 = 2x$

$x = \boxed{}$

h $3b = 42 + 16 \div 4 + 2$

$b = \boxed{}$

Good work!

96

Continued overleaf **ch** **ALGEBRA**

Name .... Class .... Date ....

2 Try these. Think carefully.

a $(54 + 56) \div (73 - 62) = x$

$x = \boxed{}$

b $(-4 + 6)18 + 134 = 5y$

$y = \boxed{}$

c $12a = (-2 - 6) + (8 \times 7)$

$a = \boxed{}$

d $b = 8^2 \times 2$

$b = \boxed{}$

e $x = 6^2 \div (33 - 3 \times 8)$

$x = \boxed{}$

f $a = 541 - (127 + 7)\, 2^2$

$a = \boxed{}$

g $(-8 + 12) + 8 \times 9 = 2y$

$y = \boxed{}$

h $16^2 \div (48 - 16) = b$

$b = \boxed{}$

i $4 + 16(53 + 18) = x$

$x = \boxed{}$

j $8^2 + 2^2 + (1056 - 841) = y$

$y = \boxed{}$

k $172 + \frac{16}{4} = 16b$

$b = \boxed{}$

l $8 \times \frac{15}{3} + 8 = 4x$

$x = \boxed{}$

m $9^2 + (14 - 2)\,\frac{360}{180} = 5a$

$a = \boxed{}$

n $6^2(156 + 84) - 460 = 4y$

$y = \boxed{}$

(ch) **ALGEBRA**

Name................................. Class..................... Date.......................

Solve equations with an unknown number on both sides.

1 Solve the problems below.

EXAMPLE $5x + 4 = x + 12$ Whatever you do to one side you must do to the other.

STEP 1

Put the x terms on the same side of the equation by subtracting x from both sides.

$$5x - x + 4 = 12$$

STEP 3

Put the number terms on the same side of the equation by subtracting 4 from both sides.

$$4x = 12 - 4 = 8$$

STEP 2

Simplify.

$$4x + 4 = 12$$

STEP 4

To find the value of x, divide both sides by 4.

$$x = 8 \div 4 = 2$$

a $7y + 2 = y + 8$

STEP 1 ....

STEP 2 ....

STEP 3 ....

STEP 4 $y = \boxed{}$

b $8b + 4 = 3b + 14$

STEP 1 ....

STEP 2 ....

STEP 3 ....

STEP 4 $b = \boxed{}$

c $12x + 4 = 9x + 13$

STEP 1 ....

STEP 2 ....

STEP 3 ....

STEP 4 $x = \boxed{}$

d $15a + 6 = 9a + 30$

STEP 1 ....

STEP 2 ....

STEP 3 ....

STEP 4 $a = \boxed{}$

Name... Class...................... Date.......................

2 Have a go at these. Think carefully about the steps.

a $34a + 1 = 18a + 17$

STEP 1 ....
STEP 2 ....
STEP 3 ....
STEP 4 $a = \boxed{}$

b $24x + 164 = 32x + 116$

STEP 1 ....
STEP 2 ....
STEP 3 ....
STEP 4 $x = \boxed{}$

c $42y + 28 = 100 + 6y$

STEP 1 ....
STEP 2 ....
STEP 3 ....
STEP 4 $y = \boxed{}$

d $4a + 82 = 16a + 22$

STEP 1 ....
STEP 2 ....
STEP 3 ....
STEP 4 $a = \boxed{}$

e $59b + 108 = 16 + 82b$

STEP 1 ....
STEP 2 ....
STEP 3 ....
STEP 4 $b = \boxed{}$

f $65x + 12 = 49x + 36$

STEP 1 ....
STEP 2 ....
STEP 3 ....
STEP 4 $x = \boxed{}$

g $3a + 27 = 9a + 12$

STEP 1 ....
STEP 2 ....
STEP 3 ....
STEP 4 $a = \boxed{}$

h $12y + 73 = 56 + 16y$

STEP 1 ....
STEP 2 ....
STEP 3 ....
STEP 4 $y = \boxed{}$

Continued overleaf (ch) **ALGEBRA**

Name.. Class....................... Date........................

---

3 Now try these. You will need to remember **BODMAS** as well.

a $(12a + 4a) + 27 = 6a + 67$

STEP 1 ....

STEP 2 ....

STEP 3 ....

STEP 4 $a = $ ☐

---

b $(4x + 2x) + 4 = 4x + 8$

STEP 1 ....

STEP 2 ....

STEP 3 ....

STEP 4 $x = $ ☐

---

c $34y + (2 + 1)4 = 76 + 2y$

STEP 1 ....

STEP 2 ....

STEP 3 ....

STEP 4 $y = $ ☐

---

d $(3 + 5)4 + 8b = 14b + 8$

STEP 1 ....

STEP 2 ....

STEP 3 ....

STEP 4 $a = $ ☐

(ch) **ALGEBRA**

ANSWERS

Page 1
1a) 5 **b)** 7 **c)** 61 **d)** 43 **e)** 113 **f)** 112
2a) 4 **b)** 19 **c)** 8 **d)** 99 **e)** 112 **f)** 195
3a) 62 **b)** 88 **c)** 61 **d)** 178 **e)** 264 **f)** 387

Page 2
1a) 54 **b)** 264 **c)** 52 **d)** 79 **e)** 182 **f)** 1253
2a) 79 **b)** 64 **c)** 203 **d)** 136 **e)** 6841 **f)** 2374
3a) 222 **b)** 280 **c)** 251 **d)** 424 **e)** 1082 **f)** 4726

Page 3
1a) Example **b)** correct combinations of 85, 33
& 52 **c)** correct combinations of 144, 24 & 168
d) correct combinations of 127, 83 & 44

Page 4
e) correct combinations of 423, 68 & 491
f) correct combinations of 423, 141 & 282
g) correct combinations of 306, 124 & 430
h) correct combinations of 588, 431 & 157

Page 5
1a) Example **b)** correct combinations of 165, 63
& 102 **c)** correct combinations of 432, 51 & 483
d) correct combinations of 364, 153 & 211

Page 6
e) correct combinations of 356, 506 & 862
f) correct combinations of 722, 336 & 386
g) correct combinations of 4723, 276 & 4999
h) correct combinations of 6032, 1382 & 4650

Page 7
1a) 56 – 23 **b)** 86 – 25 **c)** 43 + 14

Page 8
2a) 56 + 27 **b)** 83
3a) £3.50 + £5.00 **b)** £8.50

Page 9
4a) appropriate equation e.g. 82 - 65 = 17
b) appropriate equation e.g. 79 – 23 = 56
c) appropriate equation e.g. 126 – 42 = 84

Page 10
1a) 182 - 73 **b)** 473 + 189 **c)** 256 - 79

Page 11
2a) 1652 - 967 **b)** 685
3a) 1908 - 1184 **b)** 724

Page 12
4a) appropriate equation e.g. £375.50 + £50.78
= £426.28 **b)** appropriate equation e.g. 156.35
– 78.5 = 77.85 **c)** appropriate equation e.g.
87 km + 87 km = 174 km

Page 13
1a) 6 **b)** 5 **c)** 12 **d)** 6 **e)** 15 **f)** 12
2a) 27 **b)** 24 **c)** 8 **d)** 35 **e)** 12 **f)** 72
3a) 8 **b)** 42 **c)** 9 **d)** 260 **e)** 12 **f)** 98

Page 14
1a) 7 **b)** 13 **c)** 12 **d)** 18 **e)** 52 **f)** 30
2a) 96 **b)** 11 **c)** 2280 **d)** 4 **e)** 1008 **f)** 4823
3a) 31 **b)** 966 **c)** 12 **d)** 1056 **e)** 16 **f)** 1512

Page 15
1a) example **b)** correct combinations of 54, 9 & 6
c) correct combinations of 12, 8 & 96
d) correct combinations of 78, 6 & 13

Page 16
e) correct combinations of 13, 15 & 195 **f)**
correct combinations of 153, 9 & 17 **g)** correct
combinations of 23, 12 & 276 **h)** correct
combinations of 72, 8 & 9

1

ALGEBRA

ANSWERS

Page 17
1a) Example b) correct combinations of 322, 14 & 23 c) correct combinations of 14, 16 & 224 d) correct combinations of 345, 23 & 15

Page 18
e) correct combinations of 17, 26 & 442
f) correct combinations of 1352, 26 & 52
g) correct combinations of 25, 32 & 800
h) correct combinations of 1176, 56 & 21

Page19
1a) £27 ÷ 3 b) 10 x 8 c) £2.50 x 6

Page 20
2a) 9 x 12 b) 108
3a) 56 ÷ 8 b) 7

Page 21
4a) appropriate equation e.g. £12 ÷ £6 = £2
b) appropriate equation e.g. 23 x 3 = 69
c) appropriate equation e.g. 27 x 6 = 162

Page 22
1a) £21.70 ÷ 7 b) £1.50 x 56 c) 592 ÷ 16

Page23
2a) 21 x 18 b) 378
3a) 56 x 55g b) 3080g

Page 24
4a) appropriate equation e.g. 768 ÷ 12 = 64
b) appropriate equation e.g. 98 x 32 = 3136
c) appropriate equation e.g. 872 x 22 = 19184

Page 25
1a) 21 b) 19 c) 69 d) 30
2a) 30, b) 3 c) 75 d) 4
3a) 72 b) 33 c) 144 d) 9 e) 39 f) 288 g) 102 h) 8

Page 26
1a) 24 b) 63 c) 87 d) 63
2a) 135 b) 8 c) 648 d) 7
3a) 207 b) 90 c) 2430 d) 9 e) 198 f) 9720
g) 342 h) 30

Page 27
1a) 3 b) 23 c) 42 d) 88
2a) 8 b) 9 c) 12 d) 90
3a) 72 b) 48 c) 11 d) 7 e) 2 f) 79 g) 100 h) 105

Page 28
1a) 56 b) 87 c) 55 d) 346
2a) 3 b) 13 c) 6 d) 936
3a) 69 b) 1082 c) 12 d) 6615 e) 463 f) 1134
g) 2092 h) 8715

Page 29
1a) 17 b) 22 c) 6 d) 12 e) 20 f) 24 g) 9 h) 52
i) 11 j) 54 k) 16 l) 66

Page 30
2a) 8 b) 7 c) 30 d) 3 e) 12 f) 77 g) 4 h) 392
i) 19 j) 4 k) 15 l) 120

Page 31
3a) 22 b) 29 c) 4 d) 24 e) 9 f) 67 g) 4 h) 15
i) 150 j) 2 k) 222 l) 648

Page 32
4a) 9 b) 93 c) 17 d) 512 e) 8, 102 f) 89,168 g)
12, 2 h) 105, 25

Page 33
1a) 29 b) 65 c) 13 d) 108 e) 79 f) 163 g) 14
h) 46 i) 84 j) 23 k) 22.75 l) 524.5

ANSWERS

Page 34
2a) 62 b) 2394 c) 8112 d) 16 e) 33.2 f) 124.9
g) 1.25 h) 5880 i) 694.2 j) 3.5 k) 122.58 l) 336

Page 35
3a) 62 b) 199.5 c) 62 d) 31689 e) 56 f) 90.46
g) 17 h) 22.32 i) 12864 j) 35 k) 5907 l) 491.13

Page 36
4a) 34 b) 6384 c) 15.77 d) 285.6 e) 2, 93.64
f) 34.36, 953.37

Page 37
1a) 16 b) 52 c) 8 d) 114
2a) 28 b) 8 c) 26 d) 89 e) 44 f) 72 g) 105 h) 56

Page 38
3a) 11 b) 9 c) 5 d) 42 e) 6 f) 121 g) 12 h) 9
4a) 26 b) 91 c) 12 d) 96 e) 8 f) 43 g) 67 h) 12

Page 39
1a) 83 b) 18 c) 9 d) 287
2a) 105 b) 145 c) 305 d) 505 e) 239 f) 611
g) 1219 h) 718

Page 40
3a) 19 b) 15 c) 26 d) 742 e) 38 f) 40 g) 53 h) 2124
4a) 780 b) 717 c) 37 d) 19113 e) 5 f) 38.8 g) 78.6
h) 84

Page 41
1a) 11 b) 11 c) 14 d) 27
2a) 4 b) 16 c) 16 d) 9
3a) 4 b) 43 c) 14 d) 21 e) 96 f) 37

Page 42
4a) 8 b) 8 c) 5 d) 7
5a) 3 b) 30 c) 5 d) 54
6a) 4 b) 6 c) 5 d) 36 e) 28

Page 43
7a) 42 b) 96 c) 89 d) 12 e) 211 f) 12 g) 20
h) 71 i) 31 j) 10 k) 9 l) 5

Page 44
1a) 145 b) 114 c) 261 d) 180
2a) 122 b) 488 c) 272 d) 591
3a) 309 b) 457 c) 207 d) 565 e) 959 f) 1215

Page 45
4a) 12 b) 13 c) 14 d) 190
5a) 34 b) 666 c) 112 d) 351
6a) 948 b) 31 c) 21 d) 56 e) 126

Page 46
7a) 407 b) 2496 c) 832 d) 56 e) 248 f) 42
g) 426 h) 423 i) 811 j) 3 k) 3151 l) 22.3

Page 47
1a) 7cm b) 20m c) 26mm d) 72km
2a) 6km b) 12cm c) 75m d) 12mm
3a) 61m b) 53km c) 8cm d) 56m e) 5mm f) 13m

Page 48
1a) 124cm b) 195m c) 220mm d) 87km
2a) 105km b) 432cm c) 13m d) 16mm
3a) 225m b) 1210km c) 35cm d) 7776m
e) 7.1cm f) 7m

Page 49
1) 6cm 2) 10m 3) 3mm 4) 100m

ALGEBRA

ANSWERS

Page 50
1) 5mm 2) 47cm 3) 4.5m 4) 5.2km

Page 51
1a) 2,2 b) 2,1 c) 3 d) 3,1 e) triangle

Page 52
2a) A = -1,1 B = 1,1 b) 1, -1 c) 2 ,1, 0, -1,
or -2)

Page 53
1a) A = 1,1 B = 2,2 C = 5,2 b) 4,1 c) 2 d) 3,1, 4,2,
or 5,3

Page 54
2a) A = -1,1 B = 1,1 D = -2,-1 b) 2, -1 c) E = 1,2
and F = -1,2:1,0 and -1, 0:1, -1 and -1,-1 or 1,-2
and -1,-2

Page 55
1) 85° 2) 120° 3) 60° 4) 140°

Page 56
1) 59° 2) 98° 3) 45° 4) 88°

Page 57
1a) 12cm² b) 54m² c) 460mm²
2a) 3cm b) 7m c) 2cm
3a) 6cm³ B) 1m C) 6mm D) 3cm

Page 58
4a) 4cm B) 48m² C) 9mm
5a) 7cm² B) 4cm C) 4m
6a) 8cm B) 6m C) 12cm D) 26m

Page 59
7) Triangle - 44°, Quadrilateral - 85°, Square – all
90°, Pentagon - 100°, Regular Pentagon – all 108°

Page 60
1a) 104cm² b) 243m² c) 299mm²
2a) 7cm b) 42m c) 21cm
3a) 2cm B) 432m³ C) 12mm D) 9cm

Page 61
4A) 13cm B) 240m² C) 17mm
5A) 116cm² B) 120cm C) 62m
6A) 114m B) 620cm C) 132.5m D) 14.5cm

Page 62
7) Triangle – 39.5°, Equilateral Triangle - all
60°, Square – all 90°, Pentagon - 91°, Regular
Pentagon – all 108°

Page 63
8a) 1260°, 140° b) 1440°, 144 c) 8(octagon), 135°
d) 115° e) any appropriate combination = 540°

Page 64
1a) add 5: 29, 34, 39 b) add 9: 63, 81, 99 c)
subtract 11: 56, 23, 12 d) subtract 3: -7, -10, -13
e) subtract 25: 100, 25, -25

Page 65
2a) multiply by 2 or double: 16, 64, 128 b)
multiply by 4: 256, 1024, 4096 c) divide by 3: 81,
3, 1 d) divide by 2 or half: 40, 10, 5 e) multiply
by 5: 25, 125, 625 f) multiply by 2 or double: 20,
40, 80

Page 66
3a) increase, 1, add: 22, 28, 35 b) increase, 1,
subtract: 15, 10, 4 c) decrease, 1, add: 36, 42, 47
4a) increase difference by 2 and add: 17, 26, 37
b) decrease difference by 1 and subtract: 16,
10, 5

ALGEBRA

Page 67

5a) square numbers: 16, 36, 49 **b)** prime numbers 11, 19, 23

6a) add 1, subtract 2: 3, 1, 2 **b)** add 5, add 4: 26, 30, 35 **c)** subtract 5, add 2: 14, 16, 11

Page 68

1a) add 7: 75, 89, 96, 103 **b)** subtract 11: 117, 84, 73, 62 **c)** add 3: 23, 26, 32, 38 **d)** add 3: -6, 3, 6, 15 **e)** subtract 10: 55, 45, 25, 15, -5

Page 69

2a) multiply by 2 or double: 8, 16, 64, 256 **b)** multiply by 3: 24, 648, 5832 **c)** divide by 2 or half: 88, 22, 2.75 **d)** multiply by 2 or double: 0.5, 2, 8, 16 **e)** divide by 2 or half: 24, 6, 3, 0.75 **f)** divide by 2 or half: 8, 4, 1, 0.125

Page 70

3a) increase, 1, subtract: 55, 46, 41, 28 **b)** decrease, 1, add: 32, 39, 45, 50, 54

4a) increase difference by 1 and subtract: 8, 1, -10, -17 **b)** decrease difference by 1 and subtract: 48, 18, 15, 13 **c)** increase difference by 1 and subtract: 4, -6, -17, -24

Page 71

5a) square numbers: 1, 9, 25, 36, 64 **b)** prime numbers 31, 29, 23, 17, 13, 11

6a) add 4, subtract 8: 13, 9, 1, -3 **b)** double, subtract 2: 26, 52, 50, 100, 98 **c)** multiply by 3, add 2: 51, 159, 161, 483, 485

Page 72

1a) $(5 \times 3) + 1 = 16$ **b)** $(2 \times 3) + 1 = 7$ **c)** $(9 \times 3) + 1 = 28$ **d)** $(10 \times 3) + 1 = 31$ **e)** $(1 \times 3) + 1 = 4$ **f)** $(12 \times 3) + 1 = 37$

2a) $(1 \times 5) + 3 = 8$ **b)** $(3 \times 5) + 3 = 18$ **c)** $(5 \times 5) + 3 = 28$ **d)** $(8 \times 5) + 3 = 43$ **e)** $(10 \times 5) + 3 = 53$ **f)** $(14 \times 5) + 3 = 73$

Page 73

1a) $(12 \times 7) + 3 = 87$ **b)** $(15 \times 7) + 3 = 108$

2a) $1^2 \times 2 = 2$ **b)** $7^2 \times 2 = 98$ **c)** $10^2 \times 2 = 200$ **d)** $12^2 \times 2 = 288$

3a) $4^3 - 8 = 56$ **b)** $5^3 - 8 = 117$ **c)** $1^3 - 8 = -7$ **d)** $2^3 - 8 = 0$

4) $(1 \times 9) - 20 = -11$, $(2 \times 9) - 20 = -2$, $(3 \times 9) - 20 = 7$, $(4 \times 9) - 20 = 16$, $(5 \times 9) - 20 = 25$

Page 74

1a) circle round $16 - x = 7$ **b)** circle round $y = 208 \div 8$ **c)** circle round $18 = 7 + x$ **d)** circle round £y = £4 - £1.50

Page 75

2a) $a = 8 \times 4 = 32$ **b)** $b = 48 - 23 = 25$ **c)** $x = 5 \times 4 = 20$ **d)** $s = 27 \div 3 = 9$

Page 76

1a) circle round $x = 56 + 127$ **b)** circle round $252 \div 14 = y$ **c)** circle round £4.50 x 4 = b **d)** circle round $1364 + 956 = a$

Page 77

2a) $a = 1486 - 652 = 834$ **b)** $b = 876 \times 3 = 2628$ **c)** x = £24.45 ÷ 3 = £8.15 **d)** y = £2.64 + £1.25 + £15.72 = £19.61

ANSWERS

Page 78
1a) 10 **b)** 32 **c)** 6 **d)** 5 **e)** 8 **f)** 6 **g)** 40 **h)** 22
2a) 3 **b)** 2 **c)** 5 **d)** 4 **e)** 46 **f)** 72 **g)** 11 **h)** 10

Page 79
3a) 4 **b)** 49 **c)** 3 **d)** 42 **e)** 11 **f)** 33 **g)** 2 **h)** 37
4a) 36 **b)** 75 **c)** 4.8 **d)** 621

Page 80
1a) 41 **b)** 168 **c)** 6 **d)** 14 **e)** 2 **f)** 189 **g)** 52
h) 487
2a) 484 **b)** 28616 **c)** 2189 **d)** 2.5 **e)** 8089
f) 1288 **g)** 4148 **h)** 24.8

Page 81
3a) 89 **b)** 88 **c)** 9 **d)** 2460 **e)** 16 **f)** 4934 **g)** 23
h) 16.86
4a) 2 **b)** 48.439 **c)** 85.236 **d)** 390.18

Page 82
1a) 4 **b)** 4 **c)** 12 **d)** 18
2a) 20 **b)** 9 **c)** 3 **d)** 63 **e)** 8 **f)** 2

Page 83
3a) 8 **b)** 4 **c)** 39 **d)** 8
4a) 4 **b)** 6 **c)** 6 **d)** 21 **e)** 8 **f)** 22

Page 84
5a) 5 **b)** 3 **c)** 4 **d)** 4 **e)** 70 **f)** 10
6a) 6 **b)** 3 **c)** 44 **d)** 3

Page 85
1a) 37 **b)** 6 **c)** 27 **d)** 36
2a) 85 **b)** 35 **c)** 5 **d)** 1080 **e)** 211 **f)** 2

Page 86
3a) 602 **b)** 62 **c)** 1012 **d)** 317
4a) 5 **b)** 82 **c)** 13 **d)** 4509 **e)** 3 **f)** 378

Page 87
5a) 98 **b)** 214 **c)** 14 **d)** 13 **e)** 3536 **f)** 42
6a) 5.5 **b)** 232 **c)** 308 **d)** 1.5

Page 88
1a) answers from 1 x 24, 2 x 12, 3 x 8, 4 x 6 **b)**
appropriate answers **c)** appropriate answers
d) appropriate answers **e)** answers from 1 x 36,
2 x 18, 3 x 12, 4 x 9, 6 x 6 **f)** appropriate answers
g) appropriate answers **h)** appropriate answers

Page 89
2a) 1 x 28, 2 x 14, 4 x 7 **b)** 1 x 40, 2 x 20, 4 x 10,
5 x 8 **c)** appropriate answers **d)** appropriate
answers **e)** appropriate answers **f)** appropriate
answers

Page 90
3a) 1 + 5, 2 + 4, 3 + 3, 4 + 2, 5 + 1 **b)** 1 x 12, 2 x
6, 3 x 4, 4 x 3, 6 x 2, 12 x 1 **c)** 10 + 20, 11 + 19, 12
+ 18, 13 + 17, 14 + 16, 15 + 15

Page 91
1a) answers from 1 x 96, 2 x 48, 3 x 32, 4 x 24, 6 x
16, 8 x 12 **b)** appropriate answers **c)** appropriate
answers **d)** appropriate answers **e)** answers
from 1 x 110, 2 x 55, 5 x 22, 10 x 11
f) appropriate answers **g)** appropriate answers
h) appropriate answers

ANSWERS

Page 92

2a) answers from 1 x 156, 2 x 78, 3 x 52, 4 x 39, 6 x 26   **b)** 1 x 250, 2 x 125, 5 x 50, 10 x 25 **c)** appropriate answers **d)** appropriate answers **e)** appropriate answers **f)** appropriate answers

Page 93

14 + 2, 13 + 3, 12 + 4, 11 + 5, 10 + 6, 9 + 7 **b)** 1 x 20, 2 x 10, 4 x 5, 5 x 4, 10 x 2, 20 x 1 **c)** 20 + 20, 21 + 19, 22 + 18, 23 + 17, 24 + 16, 25 + 15

Page 94

1a) 13 **b)** 5 **c)** 52 **d)** 9 **e)** 3 **f)** 31 **g)** 46 **h)** 27

Page 95

2a) 20 **b)** 33 **c)** 96 **d)** 3 **e)** 8 **f)** 2 **g)** 5 **h)** 6 **i)** 11 **j)** 4 **k)** 4 **l)** 7 **m)** 11 **n)** 6

Page 96

1a) 49 **b)** 8 **c)** 122 **d)** 105 **e)** 12 **f)** 14 **g)** 177 **h)** 16

Page 97

2a) 10 **b)** 34 **c)** 4 **d)** 128 **e)** 4 **f)** 5 **g)** 38 **h)** 8 **i)** 1140 **j)** 283 **k)** 11 **l)** 12 **m)** 21 **n)** 2045

Page 98

1a) appropriate steps, **1**
b) appropriate steps, **2**
c) appropriate steps, **3**
d) appropriate steps, 4

Page 99

2a) appropriate steps, 1
b) appropriate steps, 6
c) appropriate steps, 2
d) appropriate steps, 5
e) appropriate steps, 4
f) appropriate steps, 1.5
g) appropriate steps, 2.5
h) appropriate steps, 4.25

Page 100

3a) appropriate steps, 4
b) appropriate steps, 2
c) appropriate steps, 2
d) appropriate steps, 4